AQA

GCSE (9-1)

T0187273

WORKBOOK

Geography

Practise your exam skills • Answer questions confidently • Improve your grade

Andy Owen

HODDER EDUCATION
LEARN MORE

Contents

Introduction: What is assessed on each paper?

Figure 1 shows what is assessed on each of the three exam papers. There are some options in Paper 1 and Paper 2. Make sure you know which ones you have covered.

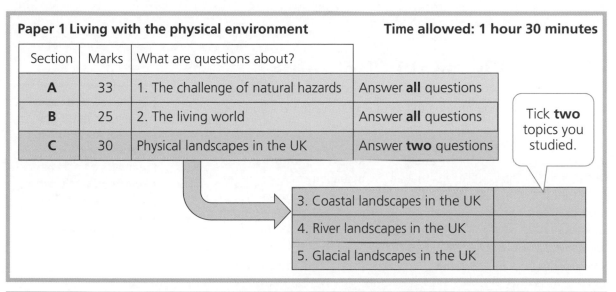

Paper 1 Living with the physical environment Time allowed: 1 hour 30 minutes

Section	Marks	What are questions about?	
A	33	1. The challenge of natural hazards	Answer **all** questions
B	25	2. The living world	Answer **all** questions
C	30	Physical landscapes in the UK	Answer **two** questions

3. Coastal landscapes in the UK	
4. River landscapes in the UK	
5. Glacial landscapes in the UK	

Tick **two** topics you studied.

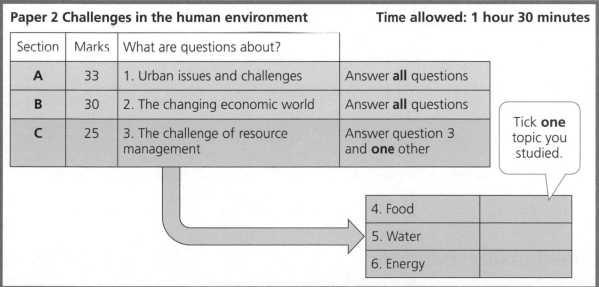

Paper 2 Challenges in the human environment Time allowed: 1 hour 30 minutes

Section	Marks	What are questions about?	
A	33	1. Urban issues and challenges	Answer **all** questions
B	30	2. The changing economic world	Answer **all** questions
C	25	3. The challenge of resource management	Answer question 3 and **one** other

4. Food	
5. Water	
6. Energy	

Tick **one** topic you studied.

Paper 3 Geographical applications Time allowed: 1 hour 30 minutes

Section	Marks	What are questions about?	
A	37	Issue evaluation	Answer **all** questions
B	39	Fieldwork	Answer **all** questions

Figure 1 What each exam paper assesses

Chapter 1: How Geography is assessed in Papers 1 and 2

This chapter is about how GCSE Geography is assessed in Papers 1 and 2. It will cover:

- what the exam questions mean
- how to tackle questions that use graphs, maps and photos
- how to answer questions worth 6 and 9 marks.

Understanding exam questions

Papers 1 and 2 have a variety of questions designed to test your ability as a geographer. It's important you understand what each question is asking you to do:

- **Command words** are words such as 'Assess' or 'Explain'. The command word tells you what you must do when you write your response. Common command words used in Paper 1 and Paper 2 are given in **Figure 1.2**.
- The **tariff** is the number of marks that are available for each question. These marks are shown at the end of the space where you put your answer. Use the number of lines printed on the exam paper as a guide to how much you should write.
- The **assessment objective (AO)** is what the examiner is looking for in your response. There are four AOs. They are described in **Figure 1.1**. Some questions assess only AO1 or AO4. These questions have a low tariff. Other questions assess a combination of AOs. These questions have 4, 6 or 9 marks. You will need to read these questions very carefully to understand what the examiner is looking for.

> In Paper 1 and Paper 2 you have about one minute for each mark. Spend about ten minutes on a 9-mark question. Don't write a lot for a 1- or 2-mark question.

Figure 1.1 The assessment objectives (AOs)

	What the examiner is looking for	Typical command word
AO1	Your ability to remember geographical facts	Describe, Give, Outline, State
AO2	Whether you understand geographical concepts and processes	Explain, Give one reason, Outline one reason
AO3	Whether you can evaluate evidence or use evidence to make a decision	Assess, Discuss, Suggest, To what extent?
AO4	Your skill when you use maps and graphs or make calculations	Describe, Calculate, Give, State

Read the question carefully

It is essential to do what the command word asks you to do. If the command is 'assess' or 'to what extent?' then you must do some evaluation or make a judgement. **Figure 1.2** lists common command words and explains what they mean.

> BUG the question! Sometimes candidates seem to write everything they know about a subject, without actually answering the question! To avoid this, **BUG** the question:
>
> **Bold** the command word.
>
> <u>Underline</u> other important instructions.
>
> *Glance* back at the question to make sure you are actually answering it!

Figure 1.2 Command words that could be used in Paper 1 and Paper 2

Tariff	Command word	What you need to do	Example
1, 2, 3	Calculate	Work out the value of something.	**Calculate** the mean shown in Figure 1. Show your working. (2 marks)
	Describe	Give a brief account of something.	**Describe** the distribution of countries shown in Figure 1. (2 marks)
	Give	Make a short, simple statement.	**Give** one reason why tropical regions have high temperatures throughout the year. (1 mark)
	Identify	Name a feature.	Using Figure 1, **identify** the landform marked X. (1 mark)
	Outline	Give a brief account of something.	**Outline** one way in which trade has had an impact on a named LIC or NEE country. (2 marks)
	State	Make a short, simple statement of fact.	**State** one physical characteristic of a tropical rainforest. (1 mark)
2–4–6	Discuss	Consider the arguments that can be debated around a geographical issue.	Using Figure 1 and your own understanding, **discuss** the issues arising from the growth of major cities in LICs or NEEs. (6 marks)
	Explain	Show your understanding by giving reasons.	**Explain** how waterfalls may change over time. (4 marks)
	Suggest	Propose a possible solution, reason or consequence. Your suggestion should be based on geographical evidence.	Using Figure 1 and your own understanding, **suggest** how large-scale agriculture can create disadvantages for the environment. (4 marks)
9	Assess	Evaluate a situation.	**Assess** the extent to which people can adapt to climate change. (9 marks)
	To what extent?	Make a judgement by weighing up the arguments for and against. Make sure you give reasons for your decision.	**To what extent** have people been successful in managing traffic congestion in a major LIC or NEE city you have studied? (9 marks)

Dealing with complex questions

Some questions seem to be very long and wordy. Don't panic. Break down the questions into bits to understand what the examiner wants you to do. In each question, look out for:

- the command – this is often (but not always) the first word in the question
- instructions to use a figure – this will be a photo, map, graph or some text in the exam paper that contains useful clues. You **must** refer to the evidence provided
- instructions to use an example or case study – you should know facts about fourteen examples and five case studies. Use details from these if the question asks for them
- whether you need to write about more than one thing – for example, a question could be about economic **and** social reasons for migration. Sometimes students do the first part (economic, in this example) and forget to do the second (social) so they don't finish the question.

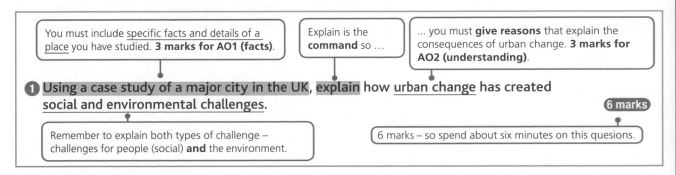

Figure 1.3 How to break down a complex question into its parts

Tackling skills questions

Papers 1 and 2 each have 19 marks that assess AO4 (skills). Most of these questions are worth 1 or 2 marks. Skills questions are about:

- using graphs
- reading maps
- doing simple calculations
- reading photos.

Using graphs

Graphs are used to present geographical data. Graph questions may ask you to:

- read a value from the graph
- complete the graph by adding a data point
- describe the shape (pattern or trend) of the graph.

Exam papers may contain a variety of graphs, including **bar charts**, **line graphs**, **climate graphs** and **scatter graphs**.

> Line graphs often show a trend. Use words that describe this trend carefully. The following words are helpful:
> - **Decreasing** if the values are going down. Add **slowly**, **steadily** or **rapidly**.
> - **Fluctuating** if the values are wobbling up and down.
> - **Increasing** if the values are going up. Add **slowly**, **steadily** or **rapidly**.

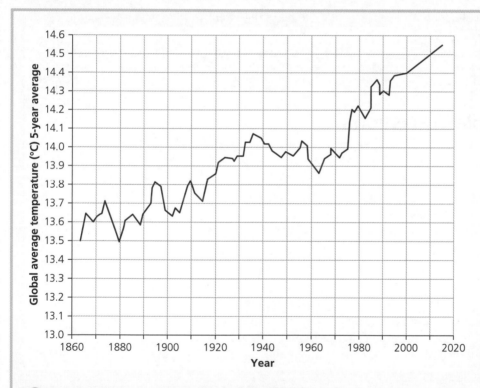

Figure 1.4 The increase in average global temperatures since 1860

1 Using **Figure 1.4**, give the temperature in:

a 1880 ....

b 1990 ....

c 2000 ....

2 Using **Figure 1.4**, which **three** of the following statements are true?

  a Temperatures increased rapidly between 1895 and 1905.

  b Temperatures decreased steadily between 1935 and 1945.

  c Temperatures increased steadily between 1915 and 1935.

  d Temperatures fluctuated between 1990 and 2015.

  e Temperatures fluctuated between 1870 and 1900.

Figure 1.5 Climate graph for Zinder, Niger

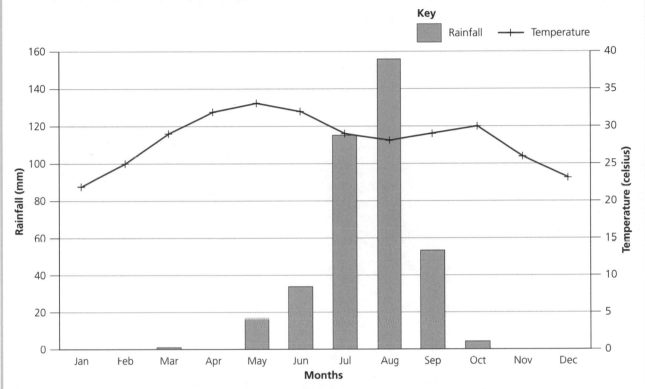

3 Using **Figure 1.5**, which **three** of the following statements are true?

  a August has 42 mm of rainfall.

  b There is a total of about 325 mm of rainfall in the three wettest months.

  c The temperature range is about 20°C.

  d The temperature range is about 12°C.

  e The minimum temperature is in August.

  f The maximum temperature is in May.

Calculate the temperature **range** by finding the difference in temperature between the hottest month and the coldest month.

Reading maps

Exam papers contain a variety of styles of maps. Some, like **Figure 1.6**, show where features are located.

Features (or data values) on a map may show a pattern. If so, the question may ask you to **describe the distribution**. In your answer, you need to describe the pattern carefully. The following words are helpful:

- **Clustered:** points on a map are concentrated into small groups.
- **Linear:** features on a map are spread out along lines.
- **Random:** features are at irregular distances from each other. There is no clear pattern.
- **Regular:** features are spaced out evenly across the map.

Other types of map are also used in exam papers, including:

- **OS maps** (see pages 48–50)
- **Choropleth maps** (see page 66)
- **Isoline maps**.

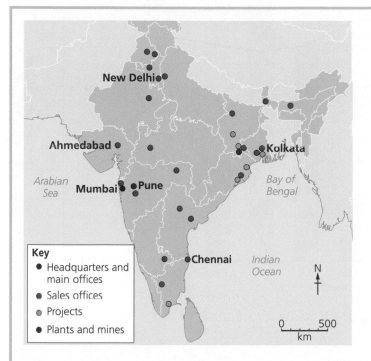

> Use words like **clustered**, **random** and **regular** to describe distribution. This is an opportunity to show the examiner your use of geographical vocabulary.

Figure 1.6 Tata steel factories and offices in India

4 Using **Figure 1.6**, which **one** of the following statements best describes the location of Kolkata?

  a Near the coast of the Bay of Bengal. ⬭

  b North of the Indian Ocean and east of New Delhi. ⬭

  c 1600 km ENE of Mumbai. ⬭

  d 1600 km WSW of Mumbai. ⬭

5 Using **Figure 1.6**, describe the location of Chennai.

....

....

> Never describe somewhere as 'near to'. It's too vague. You should always use a compass direction and distance from an important point on the map.

6 Using **Figure 1.6**, describe the distribution of Tata sales offices.

....

....

Doing simple calculations

Some questions test your ability to process geographical data by asking you to do simple calculations. The data will be presented in the exam paper as a table, graph or map. The question will begin 'Using **Figure XX**' and will use a command word such as '**calculate**'. They are simple questions which are usually worth 1 or 2 marks. You are allowed to use a calculator to find the answer. This page looks at some common questions.

Calculating averages and range

Mean: what we often think of as 'average'. To find the mean you must add up all of the values in the set of data and then divide by the number of values.

Median: to find the median value you must arrange the data in rank order. The median is the value in the middle.

Mode: the value in your set of data that occurs most frequently. It is often used with data that is sorted into categories, like the pebble sizes in **Figure 1.7**.

Range: the difference between the highest and lowest values in the set of data.

Interquartile range (IQR): the difference between values that are three-quarters and one-quarter the way through a set of data.

Study **Figure 1.7**. It shows data (the size of pebbles) collected by students at three sites along a river. **Figure 1.8** shows how you could show your working if you are asked to calculate the median or the interquartile range.

> Make sure you show how you got the answer if the question says 'Show your working here'. In a 2-mark question, you may get 1 mark by showing your working, even if your answer is wrong.

Figure 1.7 Pebble sizes (cm) at three sites along a river

Site A	14	5	3	19	8	4	2	6	15	3	12	3	7	3	9
Site B	11	4	15	2	7	1	4	3	12	8	3	1	5	2	7
Site C	6	3	1	9	3	4	1	2	6	7	2	1	4	1	5

19   15   14   **12**   9   8   7   **6**   5   4   3   **3**   3   3   2

upper quartile   median   lower quartile

> All of the numbers for **Site A** have been copied out in **rank order**

> The **median** is the value in the middle of the rank order

$IQR = 12 - 3 = 9$

> The **interquartile range** is the difference in value between the three-quarter value and the one-quarter value

Figure 1.8 How to show your working

7 Using **Figure 1.7**, calculate the median for Site B and Site C.

a   Site B ....

b   Site C ....

8 Using **Figure 1.7**, calculate the range at sites B and C.

  a Site B ....

  b Site C ....

9 Using **Figure 1.7**, calculate the interquartile range at sites B and C.

  a Site B ....

  b Site C ....

Calculating percentages

Another type of numeracy question may ask you to calculate a **percentage**. A percentage is a way of expressing part of a whole. Look at **Figure 1.9**. There are 46 megacities in the world. The whole number in this case is 46 and this is 100%. If 23 of these cities were all in the same continent, that would be 50% of the whole. To calculate the percentage of megacities that are in China, follow these steps and study **Figure 1.10**.

Step One: Divide the number of megacities in China (15) by the total number of megacities (46).

Step Two: Multiply the answer from Step One by 100.

Step Three: You can round the number up or down to the nearest whole number or to one decimal place (dp). If the number ends .49 or less, round it down. If it ends .50 or more, round it up. So 15.37 would round to 15.3% (1dp) or 15% (to the nearest whole number).

Figure 1.9 Number of world megacities (cities with a population greater than 10 million) in each continent

Continent		Number of megacities
Africa		3
Asia	China	15
	India	5
	Rest of Asia	12
Europe		3
North America		3
South America		5
Total number		**46**

Figure 1.10 How to calculate a percentage

Number of megacities in China is 15.

$$\frac{15}{46} = 0.32608$$

$$0.32608 \times 100 = 32.608$$

$$\neq 32.6 \text{ (to 1dp)}$$

10 Using **Figure 1.9**, calculate the percentage of all megacities in each of the following places. Show your answer to 1dp.

  a Rest of Asia

  b India

  c Europe

Workbook answers can be found at **www.hoddereducation.co.uk/workbookanswers**

Reading photos

Photos are often used in geography exams. The photos are not there to make the exam paper look pretty! They provide important geographical evidence, so study each one carefully.

Some questions may ask you to identify or describe a geographical feature that can be seen in the photo. They are usually worth 1 or 2 marks.

Other questions use photos to assess AO3 – whether you can think critically about evidence. You almost certainly won't have seen the photo before, but you should be able to interpret the evidence you can see in the photo by using what you have learned in your geography lessons.

Figure 1.11 A coastline in South Wales

Figure 1.12 A home on the coast of Tobago in the Caribbean

11 Using **Figure 1.11**, identify the landform marked X. `1 mark`

12 Using **Figure 1.12**, suggest how the design of the building would reduce risk during a tropical storm. `2 marks`

Student Answer

The strong concrete pillars raise the building above the height of waves (1 mark for using evidence) which reduces the risk of drowning during a storm surge (second mark for explaining how this reduces risk).

> The command is 'identify', so name the feature. The examiners will be looking for a one-word answer, in this case **headland**.
>
> 'Suggest' means you need to propose a possible answer based on what you can see. You should know that tropical storms create strong winds, large waves and a temporary rise in sea levels known as a **storm surge**.

Asking 'So what?' about photos

Other questions that use a photo could be worth 6 or 9 marks. An example of a 6-mark question is shown below. It's quite complex so the question has been broken down into its various parts. Look at **Figure 1.13**, the photo used in the exam question. You probably haven't seen this photo before, but you will have studied similar places that face similar issues. The examiner is assessing whether or not you can 'read' the geography that can be seen in the photo – using your understanding of squatter settlements to interpret the evidence shown. The photo is quite busy, so you need to study it carefully. One way to do this is to add some labels to pick out some of the key features and then to think about the impacts of these features. The labels around **Figure 1.13** have been added by the author of this book – you won't find helpful labels like this around the photos in the exam paper! Notice how each sentence is connected by the word 'so'.

> Asking yourself 'So what?' is a simple trick that will help you to improve your answers. It forces you to add explanation. Do it every time you make a simple point (like the orange labels in Figure 1.13).

These questions use a command word such as 'assess', 'discuss' or 'suggest'. The examiner is testing whether you can use evidence in the photo to build a longer response that is persuasive or evaluative.

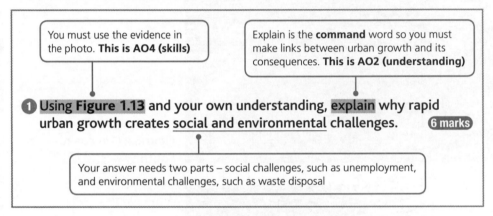

You must use the evidence in the photo. **This is AO4 (skills)**

Explain is the **command** word so you must make links between urban growth and its consequences. **This is AO2 (understanding)**

1 Using **Figure 1.13** and your own understanding, explain why rapid urban growth creates social and environmental challenges. **6 marks**

Your answer needs two parts – social challenges, such as unemployment, and environmental challenges, such as waste disposal

13 Use **three** of the labels around **Figure 1.13** to write an answer to question 1 above. Finish your answer by referring to a case study. Use the evidence from your case study in a similar way to the labels that have been added to the photo – identifying features of your city and showing how these have created challenges for poor people.

Workbook answers can be found at **www.hoddereducation.co.uk/workbookanswers**

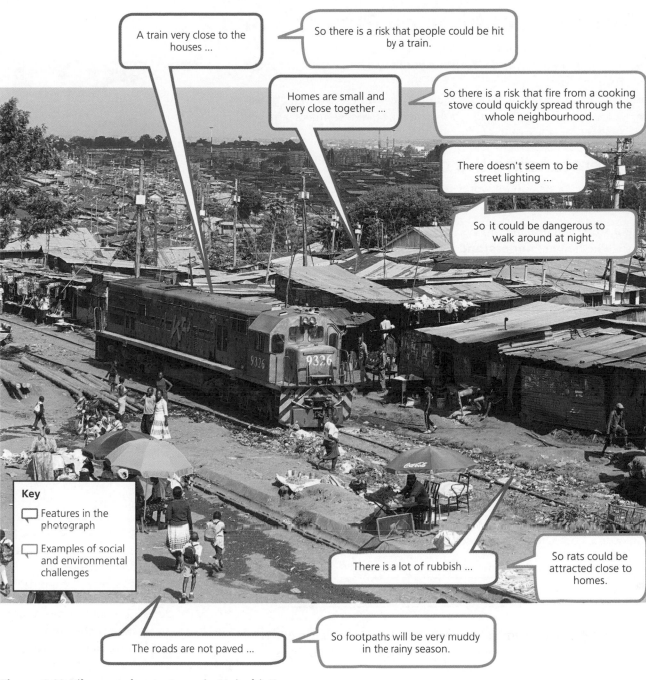

Figure 1.13 Kibera, a shanty town in Nairobi, Kenya

6-mark questions on Papers 1 and 2

Key points about 6-mark questions:

- You will need to answer **four** questions worth 6 marks in Paper 1 and another **four** in Paper 2.
- You will have six or seven minutes to answer each 6-mark question.
- A few different command words could be used, so make sure you know what they mean (**Figure 1.14**).

Understanding the question

6-mark questions can be wordy because the examiner is testing **two** assessment objectives in each question.

For a start, 3 marks are always for AO2. This is the assessment objective that tests whether you **understand** a geographical process or concept by **making links** or providing **reasons**.

However, the examiner is always looking for something else as well as AO2. It may be that you need to support your answer by referring to:

- a photo or map in the exam paper (using geographical **skills** AO4), or
- a case study or example you have learned (using your knowledge/**facts** AO1).

If so, you **must** refer to the photo/map/example or you won't finish your answer and you won't be able to get full marks. Example questions 1 and 2 below show this type of question.

Finally, the examiner may want to assess your ability to **evaluate** or **make a decision**. In this case AO2 and AO3 are assessed. Example question 3 shows this type of question.

Below, you can see examples of each of the types of 6-mark question that you might get. 6-mark questions are marked using a mark scheme with three levels – detailed, clear and basic. You won't get a tick for each point – the quality of your answer is compared to the descriptions in the mark scheme (**Figure 1.16**, page 16).

Suggest
Discuss
Do you agree? Explain your answer
Explain

Figure 1.14 Command words for 6-mark questions

You must refer to a place you have studied. **3 marks for AO1 (facts)**

1 Using a case study of a major city in a LIC or NEE, explain why urban growth has created economic opportunities and economic challenges. **6 marks**

The other marks are for explaining the links between urban growth and its consequences. **3 marks for AO2 (understanding)**

You must interpret the evidence you can see in the photo. **3 marks for AO4 (skills)**

2 Using **Figure Y** (a photo of logging in a tropical rainforest), explain how deforestation affects the environment. **6 marks**

The other marks are for explaining the links between deforestation and its consequences. **3 marks for AO2 (understanding)**

Don't panic. Read each question carefully and break it down into bits. BUG the question. Using a highlighter to pick out key words in the question is useful. Also, glancing back at the question as you write is a good idea – to make sure you are actually answering the question that has been set.

If you are told to use a figure, it will have been chosen carefully to contain evidence that might be useful in your answer. If you don't refer to it, you can't get into the top marks band.

Workbook answers can be found at **www.hoddereducation.co.uk/workbookanswers**

Explain how management can reduce risk.

3 '<u>Management strategies are effective at reducing the risk of extreme weather.</u>' Do you agree? Using an example, explain your answer.   **6 marks**

You must make a judgement and say why you have made this decision.
3 marks for AO3 (evaluation)

Don't waste time describing things if the command is 'Do you agree?'. The examiner is looking for your ability to justify your decision – so you must write a persuasive answer backed up by evidence.

14 Study the 6-mark question on page 16. Use a highlighter and underlining to break this question into bits. Make sure you identify:

a   the command word

b   the two assessment objectives

c   whether you need to write about more than one thing to fully answer the question.

15 Find one piece of evidence in **Figure 1.15** that you could use to help you answer the question. Make some notes about it in the space below.

Figure 1.15 A regenerated area of Birmingham City Centre

16 Using **Figure 1.15** and your own understanding, explain how regeneration creates social and economic opportunities.

6 marks

Figure 1.16 Mark scheme

Level	Marks	Description
3 (Detailed)	5–6	AO2 Shows detailed understanding of how regeneration is used to create social and economic opportunities. AO4 Relevant reference made to information about social and economic opportunities shown in **Figure 1.15**.
2 (Clear)	3–4	AO2 Shows clear understanding of how regeneration is used to create social and economic opportunities. AO4 Some reference made to information about social and/or economic opportunities shown in **Figure 1.15**.
1 (Basic)	1–2	AO2 Shows limited understanding of how regeneration is used to create social and/or economic opportunities. AO4 Limited or partial reference made to information about social and/or economic opportunities shown in **Figure 1.15**.

Workbook answers can be found at **www.hoddereducation.co.uk/workbookanswers**

Using connectives to develop your understanding

The examiners expect you to develop fuller answers that show a **depth** of understanding when you answer 6-mark questions. The use of **connectives** will help you to write better answers that go into more depth.

Look again at **Figure 1.13** on page 13. The points made around the photo are useful, but it would be even better if these ideas were developed even further. You can easily extend simple points by using connectives to link ideas. The easiest connective to use is 'so'. Make a point like 'There is a lot of rubbish in the photo' and ask yourself the question, 'So what?' This technique forces you to explain the consequences of the simple statement, creating longer sentences that show you fully understand something.

Study **Figure 1.17**. Read across each row. Notice how the connective 'so' has been used to link ideas.

Figure 1.17 Use the connective 'so' to turn simple statements into fully developed statements

Points	Explanations	Further explanation to add depth
The houses are made of recycled materials like corrugated tin ...	... so they will be cold at night and very hot during the day ...	... so these extremes are unhealthy for young babies or elderly people.
Houses seem to be badly built ...	... so houses built on a slope are at risk of collapse during the rainy season ...	... so people are at risk of losing their homes and all their possessions.
There is a lot of rubbish ...	... so rats could be attracted ...	... so disease carried by rats could be transmitted to the human population.
There are very few latrines ...		
The roads are not paved ...		
There isn't enough street lighting ...		

17 Study **Figure 1.17**. Use your understanding of the issues facing cities in NEEs or LICs like the one shown in the photo on page 13 to complete the elaboration for points 4, 5 and 6.

9-mark questions on Papers 1 and 2

Key points about 9-mark questions:

■ You need to answer **two** questions worth 9 marks in Paper 1 and another **two** in Paper 2:
  – one at the end of question 1
  – one at the end of question 2.
■ In each paper, **one** of these questions will have an extra 3 marks for spelling, punctuation and grammar (SPaG). This is marked on the exam paper after the question.
■ Questions will test your ability to evaluate or to make a decision (**Figure 1.18**).
■ The examiner is looking for more than one thing in your response. Questions will have:
  – 3 marks for AO1 (facts)
  – 3 marks for AO2 (understanding)
  – 3 marks for AO3 (your ability to evaluate or make a decision).

Figure 1.18 Command words for 9-mark questions

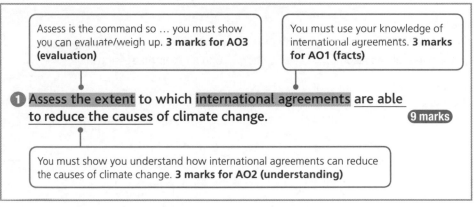

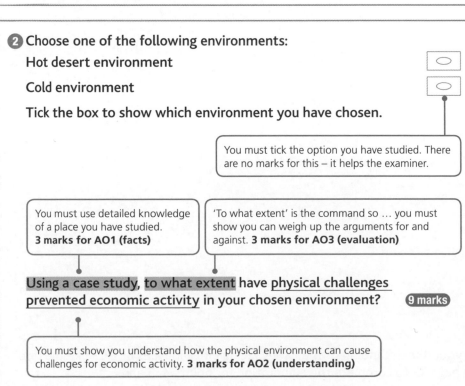

> **Some** students do the 9-mark questions first. This makes some sense – they are worth a lot of marks, so you might want to do them straight away while you are feeling alert. If you decide to do this:
> ■ allow a maximum of ten minutes (thirteen for the question with SPaG)
> ■ afterwards, work **carefully** through the paper to ensure you don't miss any questions.

Workbook answers can be found at **www.hoddereducation.co.uk/workbookanswers**

Structuring your answer

You have about ten minutes to answer the 9-mark questions (thirteen minutes for the questions with SPaG). Examiners will expect to see answers that are:
■ longer (at least eighteen lines)
■ structured – perhaps using two paragraphs and a conclusion
■ evaluative.

The question will use command words like 'Assess' or 'To what extent?', so you must give an answer that evaluates the evidence. It's a good idea to practise using a structure for your answer to this kind of question. Let's see how this can be done, using an example.

> **1** Using a case study, to what extent have physical challenges prevented economic activity in your chosen environment? **9 marks**

A good answer to this question will have three parts to it (**Figure 1.19**):
■ **An argument**. This paragraph will use evidence that supports the view. Make a point and then explain how physical challenges such as the climate or inaccessibility can prevent economic activity. Use evidence to support your argument. This technique is known as PEEL (see page 20).
■ **A counter-argument**. Use PEEL again to consider the opposing point of view, perhaps by using examples of economic activities that have been successful despite the physical challenges.
■ **A decision**. Your final paragraph should weigh up the evidence and reach a decision about whether or not physical challenges have prevented economic activity. You might come to a straightforward *yes* or *no* – a black and white decision. Alternatively, it's okay to argue for something in between. If so, use the 'washing line' technique (**see Figure 1.22**) to help you word your decision.

> Use words and phrases in your conclusion that make it really obvious to the examiner that you have answered the question.

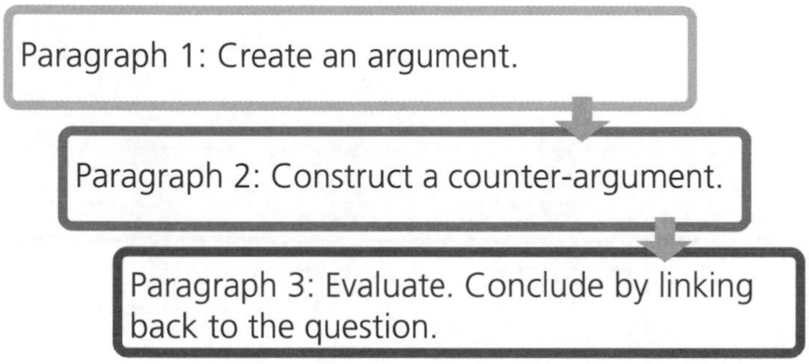

Paragraph 1: Create an argument.

Paragraph 2: Construct a counter-argument.

Paragraph 3: Evaluate. Conclude by linking back to the question.

Figure 1.19 How to construct your argument

Signposting your answer

You can use signposting to help structure your answer in a way that the examiner will find helpful and clear. Signposting is a technique that tells the reader what is coming next – like a signpost tells you where you are going. **Figure 1.20** gives a few useful signposts you can use.

To signpost an argument:	On the one hand … One view would be …
To signpost a counter-argument:	On the other hand … In comparison … Another possibility is …
To signpost your conclusion:	Overall, I think … My conclusion is …

Figure 1.20 Examples of signposting

PEEL your answer

On pages 12–13 we looked at using 'So what?' to extend and explain simple points. If you want to extend each point further, you need to PEEL it. The PEEL technique is explained in **Figure 1.21**. We have seen that a good answer to a 9-mark question will have at least two paragraphs. Each paragraph should make one point and PEEL it.

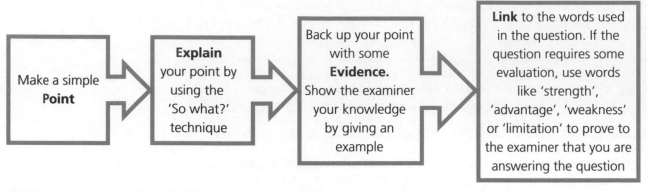

Figure 1.21 Use the PEEL technique in each paragraph

Even if you fully agree or completely disagree, you should **always** present both sides of an argument and then make a decision.

Dealing with 'to what extent?'

You **must** state whether you agree with the statement or not. You may fully agree or disagree with the statement. It's also possible that you only partially agree with it. Either way, it doesn't really matter because the examiner isn't looking for a particular answer. It's the way that you use the evidence to support your decision that is important. **Figure 1.22** gives you some helpful phrases to use in your answer.

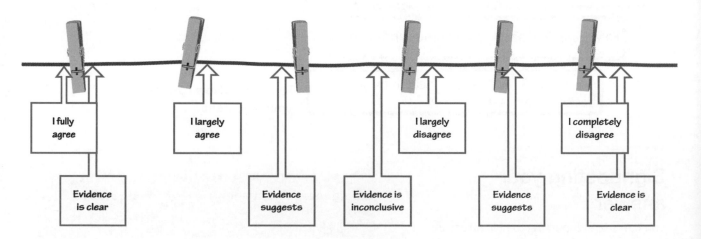

Figure 1.22 Use a 'washing line' to help you state your decision

Workbook answers can be found at **www.hoddereducation.co.uk/workbookanswers**

How 9-mark questions are marked

9-mark questions are marked using a mark scheme with three levels – detailed, clear and basic. You **won't** get a tick for each point – the quality of your answer is compared to the descriptions in the mark scheme. Study the example in **Figure 1.23**. It could be used with the following question.

> **18** Assess the extent to which international agreements are able to reduce the causes of climate change.
>
> **9 marks**

Figure 1.23 Mark scheme

Level	Marks	Description
3 (Detailed)	7–9	AO1 Detailed knowledge of international agreements. AO2 Thorough geographical understanding of how agreements are able to reduce the causes of climate change. AO3 Thorough evaluation of the extent to which agreements are able to reduce the causes of climate change.
2 (Clear)	4–6	AO1 Clear knowledge of international agreements. AO2 Some geographical understanding of how agreements are able to reduce the causes of climate change. AO3 Reasonable evaluation of the extent to which agreements are able to reduce the causes of climate change.
1 (Basic)	1–3	AO1 Limited knowledge of international agreements. AO2 Slight geographical understanding of how agreements are able to reduce the causes of climate change. AO3 Limited evaluation of the extent to which agreements are able to reduce the causes of climate change.

From the mark scheme you can see that your answer needs a combination of facts, understanding and evaluation. The sample answer below deals with each assessment objective separately so you can see the different things that the examiner might be looking for when they are using the mark scheme.

Sample answer

Facts (AO1)

The United Nations Framework Convention on Climate Change (UNFCCC) met in Paris in 2015 and 195 countries signed the Paris Agreement. Its aim is to keep the increase in global temperature to less than 2°C above levels that existed before the industrial age.

Understanding (AO2)

The agreement works because individual governments pledge to reduce emissions of greenhouse gases that trap heat in the atmosphere. For example, by pledging to stop burning coal to generate electricity and switching to more renewables, a country will emit less CO_2 so less heat will be trapped in the atmosphere.

Evaluation (AO3)

It is difficult for international agreements to make actual reductions in CO_2 emissions because there is nothing to force countries to stick to their pledges. International agreements will work only if enough countries that emit large quantities of greenhouse gases, such as the USA, India or China, actually keep to the targets. Politicians may fail to do this because of pressure from voters. Scientists think that most industrial countries are failing to meet their own targets.

Spelling, punctuation and grammar

The accuracy of your spelling, punctuation and grammar will be marked in **one** 9-mark question in **each** exam (**Figure 1.24**). That means there are 12 marks in total and you should spend about thirteen minutes answering this question. The examiner will be looking for:

- **spelling** – be especially careful with the spelling of geographical terms. The examiner will also look for common mistakes such as their/there or where/wear
- **punctuation** – make sure you get the basics right. Check that sentences and place names start with a capital letter. Keep sentences short and end them with a full stop. Use commas in lists
- **grammar** – don't write in bullet points. Organise your answer in paragraphs. The examiner may also look for common mistakes such as could of/could have.

Figure 1.24 The mark scheme for spelling, punctuation and grammar

Description	Marks
Higher performance ■ Learners spell and punctuate with consistent accuracy. ■ Learners use rules of grammar with effective control of meaning overall. ■ Learners use a wide range of specialist terms as appropriate.	3
Intermediate performance ■ Learners spell and punctuate with considerable accuracy. ■ Learners use rules of grammar with general control of meaning overall. ■ Learners use a good range of specialist terms as appropriate.	2
Threshold performance ■ Learners spell and punctuate with reasonable accuracy. ■ Learners use rules of grammar with some control of meaning and any errors do not significantly hinder meaning overall. ■ Learners use a limited range of specialist terms as appropriate.	1
No marks awarded ■ The learner writes nothing. ■ The learner's response does not relate to the question.	

Student response 1

The governments of lots of countries met in Kyoto (Japan) and Paris (France) to decide what to do about climate change. They set targets to reduce the amount of CO_2 and other greenhouse gases they emit so that less heat is trapped in the atmosphere. I don't think that international agreements can prevent climate change because some countries like India, China and the USA emit huge amounts of CO_2 because they are industrial countries and they have very large populations. Countries like Tuvalu in the Pacific hardly emit any CO_2 but they will be flooded if sea levels rise. That's why international agreements are important.

Student response 2

International agreements were made by 195 countries in Paris in 2015 to keep the increase in temperature down to an increase of only 2°C compared to what it used to be in the past so that no extra heat is trapped in the atmosphere and so ice caps stop melting causing coastal flooding and so much extreme weather. International agreements are important because they force governments to act for example France has promised to stop burning coal to make energy and also they will stop making petrol engine cars so that vehicles have much lower emissions which will slow down the rate of climate change. On the other hand agreements are not enough on their own we could of been doing things for ages to prevent the worst affects of climate change by doing things like planting more trees building sea walls improving weather forecasts etc.

19. Read student response 1. Annotate the response to show ways that you could improve:

   a   the facts about international agreements

   b   the level of evaluation.

20. Read student response 2.

   a   Why is it a better answer than response 1?

   b   How could it be improved?

   c   Identify and correct three silly SPaG mistakes.

Chapter 2: Preparing for Paper 1

Tectonic hazards

Tectonic hazards are caused by the movement of the plates that make up the Earth's crust. A small number of key geographical terms could be used in exam questions. Learn them carefully.

| plate margins | constructive | destructive | conservative | tectonic plates |

1 Match **five** key terms to the correct definition below. **Two** terms are not needed.

| plate margins | transnational | constructive | destructive |
| conservative | tectonic plates | seismometer | |

Term	Definition
	Rigid sections of the Earth's crust.
	Places where the Earth's plates meet each other.
	Plate margins where plates are moving towards one another.
	Plate margins where plates are moving away from one another.
	Plate margins where plates are sliding past one another.

Figure 2.1 The Earth's plate margins

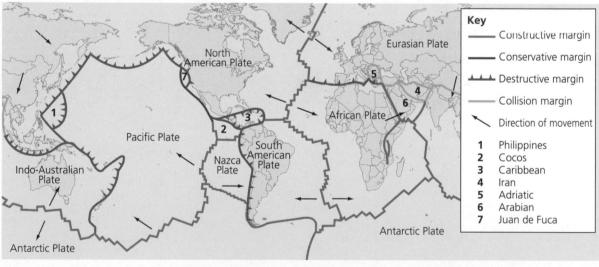

2 Using **Figure 2.1**, which **three** of the following statements are true?

a The Nazca Plate is being destroyed beneath the South American Plate.

b There is a conservative plate margin on the eastern side of the North American Plate.

c The western Pacific Plate is bordered by destructive plate margins.

d There is a destructive plate margin between the South American Plate and the African Plate.

e There is a constructive plate margin between the North American Plate and the Eurasian Plate.

Workbook answers can be found at **www.hoddereducation.co.uk/workbookanswers**

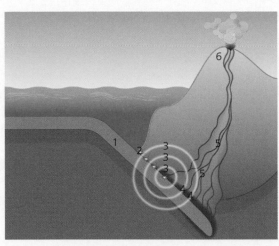

Figure 2.2 The process of subduction occurs at destructive plate boundaries

3 Using the labels in **Figure 2.2**, number each step in the following table to explain what happens as the Nazca Plate meets the South American Plate.

Number	Labels for Figure 2.2
	The magma reaches the surface, causing a volcanic eruption
	There is friction between the oceanic and continental plates
	Magma rises through the continental crust
	The dense Nazca Plate is pulled into the mantle
	The heat and friction cause oceanic crust and ocean floor sediment to melt
	The friction is overcome, causing an earthquake

Effects of tectonic hazards

Tectonic hazards have immediate effects on the environment and on people. These are **primary effects**. These effects can trigger further problems which may continue for weeks or months. These are **secondary effects**.

4 Study the list of possible effects of a volcanic eruption or earthquake. Decide whether these are primary effects or secondary effects by writing each letter into the correct part of the figure.

A Volcanic gases   **B** Lahars   **C** Lava flows   **D** Refugees

E Food prices rise   **F** Pyroclastic flows   **G** Landslides   **H** Ash clouds

I Travel disruption   **J** Crops die   **K** People injured   **L** Buildings collapse

M Shortage of water   **N** Disease spreads

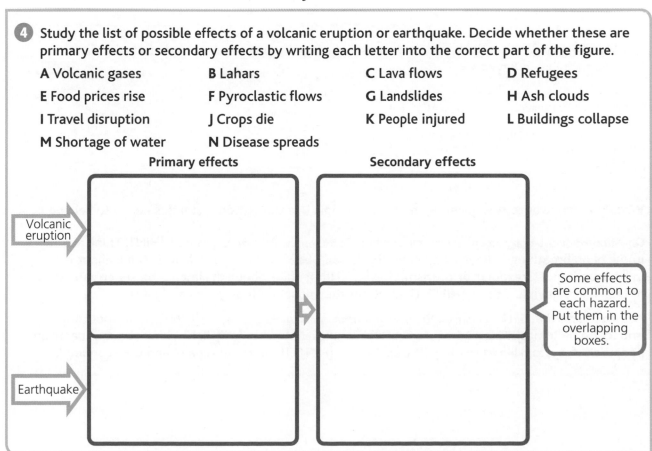

Study **Figure 2.3**. It shows a constructive plate margin that runs through the middle of Iceland. Iceland is a wealthy country.

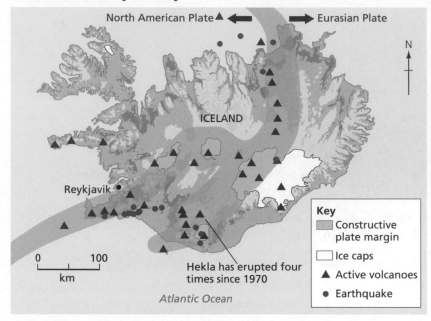

Figure 2.3 Iceland's plate margins

⑤ Using **Figure 2.3**, which **one** of the following statements best describes the distribution of tectonic hazards?

a   Regularly spaced along the constructive plate margin.

b   Randomly spread throughout Iceland.

c   Mainly arranged in a linear pattern along the constructive plate margin, with some clustered to the southeast of Reykjavik.

⑥ Using **Figure 2.3** and your own understanding, suggest how tectonic hazards could affect people living in Reykjavik. **6 marks**

You should refer to map evidence to support your answer. For example, you could work out the direction of the wind that would carry ash from Hekla over Reykjavik. **3 marks for AO3 (making judgements)**

Show you understand how tectonic activity creates dangerous hazards for people. **3 marks for AO2 (understanding)**

Write your answer to this question in your notebook. Use the sentence starters below to help you.

One hazard created by volcanic eruption is ash. If the wind was blowing … so … [HINT: Think about impacts on breathing in Reykjavik]. *Also, flights could be delayed … so …* [HINT: Think about the impact on imports/exports or tourism] *… so …* [HINT: Think about the impact on supermarkets or hotels in Reykjavik] *… so …* [HINT: Think about the impact on food prices and jobs].

Another hazard would be the ground shaking during earthquakes … so … [HINT: Think about the impact on buildings]. *This might have short-term impacts such as …* [HINT: Think about impacts on people's safety] *and long-term impacts such as …* [HINT: Think about repairs and the economy].

Managing tectonic hazards

Many people continue to live in areas at risk of tectonic hazards because they cannot afford to move or they think that the risks are low. We can reduce the risks of tectonic hazards through **monitoring**, **prediction**, **protection** and **planning**.

7 Match the **four** key terms to the examples below. You may use some terms more than once.

protection   monitoring   planning   prediction

Term	Example
	Making regular measurements of gases emitted from a volcano.
	Using data (like earth tremors) to estimate when a volcano might erupt.
	Using seismometers to measure earth tremors.
	Designing buildings so that they flex rather than collapse during an earthquake.
	Using buoys in the ocean to measure the height of a tsunami wave.
	All emergency services know what to do in the event of an earthquake.

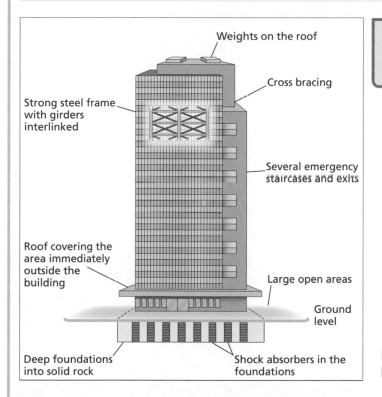

Use the 'So what?' technique (pages 12–13) to explain each point. One example has been done for you.

Figure 2.4 An earthquake-resistant building

8 Choose **two** of the labels on **Figure 2.4**. For each of these, explain how the design feature would reduce the risk. One example has been done for you.

Feature 1
A roof covers the area around the building so people are not injured by broken glass if it falls from windows.

Feature 2

....

....

Feature 3

....

....

9 Using **Figure 2.4** and your own understanding, suggest how useful monitoring and protection can be in reducing tectonic risks.

6 marks

By referring to **Figure 2.4** you can explain how useful protection can be.

The command means you must make a decision – is it useful or not in reducing risk? **3 marks for AO3 (decision making)**

Use an example of monitoring to explain how it works. **3 marks for AO2 (understanding)**

Workbook answers can be found at **www.hoddereducation.co.uk/workbookanswers**

Examples of tectonic hazards

How much damage is caused by an earthquake or eruption depends partly on the wealth of the area affected. Wealth also affects how well people are able to respond with evacuation, aid or rebuilding programmes. Use the table below to summarise **two** examples. Use bullet points.

	Richer place	Poorer place
Place name		
Primary effects of the hazard	1. 2.	1. 2.
Secondary effects of the hazard	1. 2.	1. 2.
One immediate response		
One long-term response		

List specific facts

E.g. rescuing people or providing shelter

E.g. rebuilding

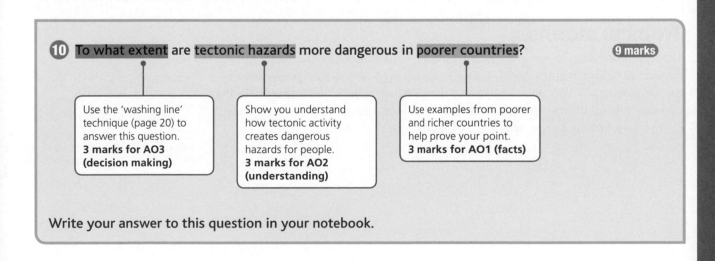

10 **To what extent** are **tectonic hazards** more dangerous in **poorer countries?**　　**9 marks**

Use the 'washing line' technique (page 20) to answer this question.
3 marks for AO3 (decision making)

Show you understand how tectonic activity creates dangerous hazards for people.
3 marks for AO2 (understanding)

Use examples from poorer and richer countries to help prove your point.
3 marks for AO1 (facts)

Write your answer to this question in your notebook.

Weather hazards

Atmospheric circulation

The movement of air in our atmosphere is driven by the Sun warming the Earth. The equatorial regions are the hottest part of the Earth because the Sun is overhead. Excess heat at the Equator triggers atmospheric circulation, causing warm air to spread outwards towards the Tropics and creating **trade winds**.

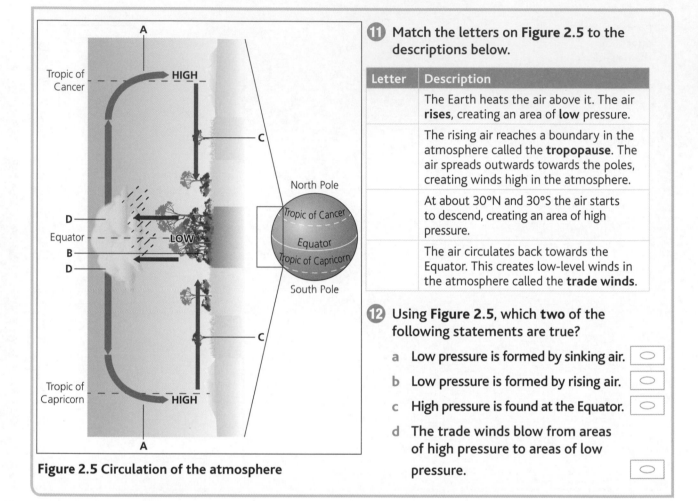

Figure 2.5 Circulation of the atmosphere

⑪ Match the letters on **Figure 2.5** to the descriptions below.

Letter	Description
	The Earth heats the air above it. The air **rises**, creating an area of **low** pressure.
	The rising air reaches a boundary in the atmosphere called the **tropopause**. The air spreads outwards towards the poles, creating winds high in the atmosphere.
	At about 30°N and 30°S the air starts to descend, creating an area of high pressure.
	The air circulates back towards the Equator. This creates low-level winds in the atmosphere called the **trade winds**.

⑫ Using **Figure 2.5**, which **two** of the following statements are true?

a Low pressure is formed by sinking air. ⬭

b Low pressure is formed by rising air. ⬭

c High pressure is found at the Equator. ⬭

d The trade winds blow from areas of high pressure to areas of low pressure. ⬭

Tropical storms

Tropical storms begin when air is warmed over warm oceans (the water has to be over 27°C for several weeks). This creates an area of very **low pressure** in the atmosphere. Tropical storms create very **strong winds** and **heavy rain** that can cause **landslides**. The low air pressure causes a temporary rise in sea levels known as a **storm surge**, which may **flood** coastal areas.

Use evidence in the photo to make a point, then use 'So what?' to explain it

Strong winds damage communications so emergency services may be delayed. Power lines are damaged so hospitals have power cuts

Flood water is polluted with sewage so disease spreads

Figure 2.6 The effects of a tropical storm in the Philippines

You must use evidence in the photo, e.g. the flooding. **3 marks for AO4 (skills)**

This is the command so you must give reasons …

13 **Using Figure 2.6** and your own understanding, explain why tropical storms need immediate and long-term responses.

6 marks

Structure your answer around these two types of response.

… and show your understanding of why different types of responses are needed. **3 marks for AO2 (understanding)**

Write two short paragraphs – one for immediate responses (e.g. clean water and medicines) and one for long-term responses (e.g. repairs to infrastructure and better flood warning systems).

Use the PEEL technique – make a point and explain it using evidence.

Immediate responses	Long-term responses

An example of a tropical storm

You will have studied an example of a tropical storm. You need to remember when it happened, the places it affected, how they were affected and how people responded to this hazard. Use the table below to summarise your example.

Name of the tropical storm	
When it happened	
Place(s) affected	
Effects on people/the economy	Effects on the environment
	E.g. on coastlines
One immediate response	
	E.g. shelter, medical help or food aid
One long-term response	
	E.g. repairs to flood defences or better forecasts

Managing weather hazards

We can reduce the risk of weather hazards through monitoring, prediction, protection and planning.

14 Match **four** key terms to the examples below. You may use each term more than once.

protection　　　　monitoring　　　　prediction　　　　planning

Term	Example
	Build storm shelters on stilts so they are above the storm surge.
	Use weather data to warn people when a tropical storm is approaching.
	Refuse permission for new homes in areas at risk of coastal floods.
	Take regular readings of water temperature, air temperature and air pressure.
	Launch weather satellites.
	Strengthen river embankments so rivers hold more water.
	Build coastal defences to prevent storm surges from flooding low-lying coastal areas.
	Make sure local people know what to do and where to go if there is a flood.
	Train fire crew and police so they know what to do after a storm.

Workbook answers can be found at **www.hoddereducation.co.uk/workbookanswers**

Extreme weather in the UK

Extreme weather in the UK can be caused by:

- **low pressure** – causing storms, wind, heavy rainfall and flooding
- **high pressure** – causing drought and heat waves in the summer.

An example of extreme weather in the UK

You will have studied an example of a recent extreme weather event in the UK. You need to remember when it happened, the places it affected, how they were affected and how people responded to this hazard. Use the table below to summarise your example.

What kind of extreme weather	
When it happened	
Place(s) affected	
Causes	
Impacts: Social Economic Environmental	
How management strategies can reduce risk	

> Use bullet points to list key facts and figures. Be specific

> E.g. repairs to flood defences or better forecasts

15 'Strategies to manage extreme weather in the UK are effective in reducing risk.' Do you agree? Using an example, explain your answer. `6 marks`

Use evidence from an example to support your argument. **2 marks for AO1 (facts)**

Explain how prediction, monitoring or planning can reduce risks. **2 marks for AO2 (understanding)**

You must make a decision – is it possible to reduce the risks caused by extreme weather or not? **2 marks for AO3 (decision making)**

Point – outline one strategy; for example, weather forecasts (prediction)

Explain how the strategy reduces risk

Give **evidence** (facts) using the example you studied

Evaluate and **link** by stating whether the strategy was effective – did it reduce risk or not?

Climate change

Evidence of climate change

Evidence of climate change comes from:

- **ice cores** (which contain CO_2 trapped in the ice)
- **temperature records** (since about 1850)
- **pollen analysis** – pollen is found in peat bogs. It tells us which plants were growing at the time, which tells us whether the climate was warm or cold
- **tree rings** (because rings are thicker when the climate is warmer and wetter).

Figure 2.7 How the Earth's temperature has varied in the last 400,000 years

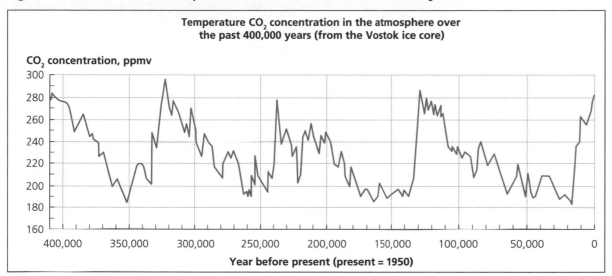

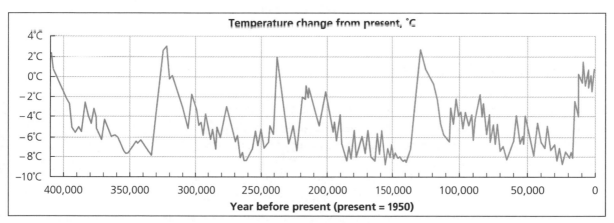

16 Using **Figure 2.7**, which **three** of the following statements are true?

a The highest concentration of CO_2 was about 275,000 years before the present.

b The peaks in temperature are **interglacial** periods. They occur when CO_2 levels are high.

c The troughs in temperature are **glacial** periods. They occur when CO_2 levels are lower than today.

d 150,000 years ago the CO_2 was at a concentration of 280 ppmv.

e 150,000 years ago the temperature was about 8°C colder than at present.

Possible causes of climate change

Some causes of climate change are natural. Factors such as changes in the Earth's orbit, volcanic eruptions and variations in solar output caused climate to change over the **Quaternary Period** (the past 2.6 million years). Human activities such as burning fossil fuels, agriculture and deforestation have caused more climate change, especially over the past 200 years.

17 Match **four** causes of climate change to the explanations below. **Tick** whether they are natural or human.

Volcanic eruptions   Agriculture   Orbital changes   Deforestation

Cause	Explanation	Natural	Human
	The path taken by the Earth as it travels around the Sun varies over a 96,000-year cycle. This means that the amount of solar energy received by the Earth varies over time.		
	Volcanoes emit SO_2, which can cool the atmosphere by blocking the amount of solar energy received by the Earth.		
	Plants take CO_2 from the atmosphere by photosynthesis and store it. When trees are chopped down and burned, this store of CO_2 is released into the atmosphere.		
	Rice farming, dairy farms and beef farms produce a lot of methane, which is a powerful greenhouse gas.		

Managing climate change

We can manage climate change by:
- **reducing causes**, for example by using renewable energy, planting trees to store carbon or capturing CO_2. This is known as **mitigation**. Politicians from different countries can make **international agreements** about reducing the amount of greenhouse gases that are emitted
- **responding to the changing climate**, for example by changing how we produce food and manage water supplies or by building better flood defences. This is known as **adaptation**.

18 Assess the extent to which international agreements are able to reduce the causes of climate change.

9 marks

Weigh up the argument for and against. **3 marks for AO3 (evaluation)**

Give details about international agreements. **3 marks for AO1 (facts)**

Show you understand how international agreements can reduce the causes of climate change. **3 marks for AO2 (understanding)**

Answer this question in three short paragraphs. First, create an argument. Second, create a counter-argument. Finally use the 'washing line' technique.

On the one hand ...

Use PEEL to make an argument that international agreements can be effective

On the other hand ...

Create a counter-argument. Use evidence to show why international agreements can fail to reduce the causes of climate change

I fully/largely/partially agree/disagree ...

Link back to the question. State whether international agreements are effective or not. Use the 'washing line' technique

Ecosystems

Plants in an ecosystem get their energy from the Sun in a process called **photosynthesis**. They use this energy to produce food in the form of leaves, seeds or fruit – so plants are **producers** in the ecosystem. In a UK woodland, the leaves, seeds and fruit are eaten by caterpillars, insects, birds or mice. These are **consumers**. Energy is recycled in an ecosystem through **nutrient cycling** – see **Figure 2.8**.

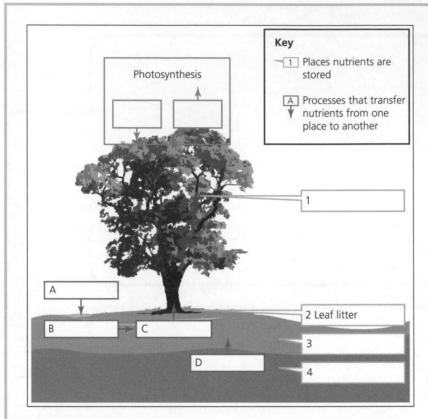

Figure 2.8 Nutrient cycling in a UK woodland

19 Complete the orange boxes in **Figure 2.8** to show the gases transferred by photosynthesis.

20 Match the following labels to the correct letter or number on the diagram to explain nutrient cycling.

Letter/number	Explanation
	Nutrients are stored in rocks.
	Weathering breaks down rocks and releases nutrients into the soil.
	Nutrients are stored in the soil.
	Nutrients are stored in the cells of the tree.
	Dead leaves and branches fall from the tree.
	Nutrients are stored temporarily in the dead leaves and branches on the forest floor.
	Beetles and earthworms break down the dead leaves. Bacteria and fungi (**decomposers**) release the nutrients into the soil.
	Water dissolves the nutrients. The tree takes in the water and nutrients through its roots.

An example of a small-scale UK ecosystem

You will have studied an example of a small-scale UK ecosystem such as a woodland, hedgerow or sand dune. Use the table below to summarise key facts and figures of your example.

Type of ecosystem	
Examples of producers	
Examples of consumers	
Example of a food chain	

Tropical rainforests

Most tropical rainforests grow within 10 degrees of the Equator. The climate is:

- hot every month, usually between 23°C and 30°C
- wet for most months of the year – total rainfall is between 2000 mm and 4000 mm per year.

The plants that grow in tropical rainforests are adapted to this climate.

Figure 2.9 Tropical rainforest plants

21 a Circle the correct word from the *options* in each description below.

b Match the descriptions to the labels on **Figure 2.9**.

Label	How plants are adapted
	Leaves are *waxy/porous* and have a drip-tip so that rain water runs off quickly, preventing rot.
	Plants grow *slowly/quickly* to reach the *nutrients/sunlight* above.
	The forest floor is *shaded/brightly lit* so plants on the forest floor have *small/large* leaves so they can catch enough sunlight.
	The upper leaves form an umbrella-like *canopy/emergent* layer. This prevents a lot of light reaching the forest floor.

Deforestation

Tropical rainforests contain a lot of useful resources, including timber, minerals (e.g. iron ore) and land for agriculture. Rainforests are also cut down to make space for energy schemes (e.g. hydro-electricity projects), new roads and new towns. Cutting down trees is called **deforestation**. Deforestation creates wealth (an economic impact) but has negative impacts for wildlife (an environmental impact).

A case study of a tropical rainforest

You will have studied a case study of the deforestation of a tropical rainforest. Use the table below to summarise the key facts and figures.

Location of the rainforest	
Facts about **three** causes of deforestation	1. 2. 3.
The impacts of deforestation on ...	1. Economic development ... 2. Soil erosion ... 3. Contribution to climate change ...

E.g. mineral extraction, farming, or logging

Use bullet points to list key facts and figures. Be specific

---

22 Explain how deforestation creates various impacts:

   a Link the **points** about deforestation on the left to the **explanations** on the right. One example has been done for you.

   b Use a highlighter to show how each explanation has been extended further.

   c Which of these impacts is negative and which is positive?

Positive impacts .... Negative impacts ....

Points	Explanations
1 The tropical rainforest stores a lot of carbon ...	... so tropical countries are able to create wealth by exporting food crops leading to a better balance of trade.
2 When trees are cut down the canopy is destroyed ...	... so when it is burned CO_2 is released into the atmosphere causing climate change.
3 Large firms create commercial farms growing palm oil or oranges ...	... so interception is reduced leading to soil erosion and a reduction in soil fertility.
4 Large areas of rainforest are replaced by farms growing one crop, such as soy ...	... so the rainforest becomes more accessible, which means that more people move in looking for work.
5 Deforestation allows minerals such as iron ore to be extracted ...	... so the rich variety of plants is replaced by just one species of plant, which means much less biodiversity.
6 Roads are built through the forest to remove logs ...	... so well-paid jobs are created in mining and engineering, which leads to a demand for better training and research.

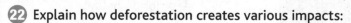

The 'So what?' technique has been used to explain each point (highlighted in yellow). Notice how the sentence has been extended again to add further explanation (highlighted in green).

Workbook answers can be found at **www.hoddereducation.co.uk/workbookanswers**

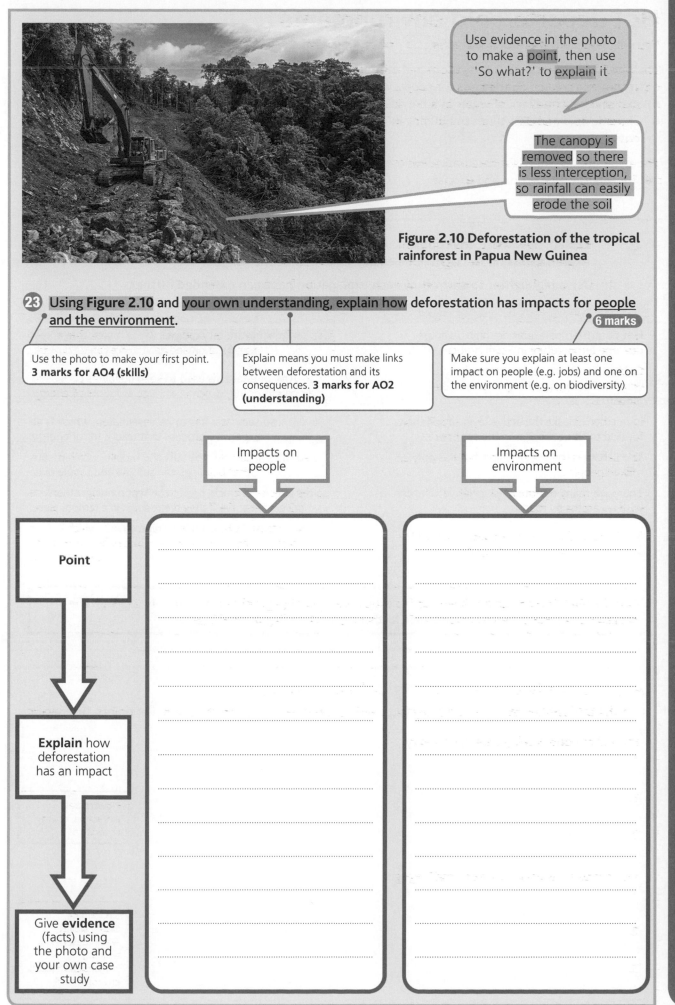

Use evidence in the photo to make a point, then use 'So what?' to explain it

The canopy is removed so there is less interception, so rainfall can easily erode the soil

Figure 2.10 Deforestation of the tropical rainforest in Papua New Guinea

23 Using **Figure 2.10** and your own understanding, explain how deforestation has impacts for people and the environment.

6 marks

Use the photo to make your first point. **3 marks for AO4 (skills)**

Explain means you must make links between deforestation and its consequences. **3 marks for AO2 (understanding)**

Make sure you explain at least one impact on people (e.g. jobs) and one on the environment (e.g. on biodiversity)

Impacts on people

Impacts on environment

Point

Explain how deforestation has an impact

Give **evidence** (facts) using the photo and your own case study

Hot deserts/cold environments

You will have studied **either** hot deserts ⬭ **or** cold environments ⬭ (tick one).

The physical environment (**ecosystem** and **climate**) of hot deserts and cold environments creates **challenges** for people. Challenges include:

- extreme heat and lack of water in a hot desert
- extreme cold and difficulties in building infrastructure (such as roads) in cold environments.

There are opportunities for economic development in both environments, such as mining, energy production and tourism.

 a Explain how these environments create challenges by linking the **points** on the left to the **explanations** on the right. One example has been done for you.

b Use a highlighter to show how each explanation has been extended further.

Points	Explanations
Hot desert temperatures are high. Average daytime temperature in Saudi Arabia is 45°C …	… so settlements are very remote, which means that access to education and health care can be difficult.
Cold environments have extremely cold winters. The average temperature in Barrow in January is −30°C …	… so 70% of energy is used to keep buildings cool using air conditioning, which uses huge amounts of energy.
In northern Alaska the active layer of soil thaws for two to three months in the summer …	… so seawater has to be desalinated, which is an expensive process that uses a lot of energy.
The Sahara is the size of China, but has only 2.5 million people …	… so the soil can become soft and mobile, which means that buildings can subside and topple over.
There are many remote Alaska Native villages with no access by road …	… so the engines of trucks need to be kept running, otherwise drivers are at risk if they break down in a remote place.
Saudi Arabia does not have enough rainfall to provide a sustainable water supply …	… so at least 75 have no law enforcement, which means that it can take over 24 hours before police come to an emergency call.

> The 'So what?' technique has been used to explain each point (highlighted in yellow). Notice how the sentence has been extended again to add further explanation (highlighted in green).

A case study of hot deserts/cold environments

Use the table below to summarise key facts and figures of your case study. Use bullet points. Be specific.

Facts about **one** development opportunity:

1.

2. *E.g. mineral extraction, energy or tourism*

3.

What makes this environment so challenging:

1.

2. *E.g. climate, inaccessibility, water (deserts), building infrastructure (cold environments)*

3.

Workbook answers can be found at **www.hoddereducation.co.uk/workbookanswers**

Choose one of the following environments:

Hot desert environment ⬭

Cold environment ⬭

Tick the box to show which environment you have chosen.

25 Using a case study, to what extent have the physical challenges prevented economic activity in your chosen environment? **9 marks**

You must use your knowledge of a case study.
3 marks for AO1 (facts)

Use the 'washing line' technique (page 20) to answer this question.
3 marks for AO3 (decision making)

Show you understand how climate or remote locations create challenges for business.
3 marks for AO2 (understanding)

Answer this question in three short paragraphs. First, create an argument. Second, create a counter-argument. Finally, use the 'washing line' technique.

On the one hand ...

Use **PEEL** to make an argument that physical challenges prevent economic activity

On the other hand ...

Create a counter-argument. Use evidence to show how businesses can overcome problems such as extremes of temperature

I fully/largely/partially agree/disagree ...

Link back to the question. State whether physical challenges prevent economic activity or not. Use the 'washing line' technique (page 20)

Physical landscapes in the UK

You will have studied **two** types of UK physical landscape. Tick **two** examples you have studied:

Coastal landscapes ⬭   River landscapes ⬭   Glacial landscapes ⬭

Landforms

The processes of **erosion**, **transportation** and **deposition** cause landforms to change over time. Make sure you can:

- use the correct technical terms
- explain how each process causes the landform to change.

26 Choose the two landscapes you have studied.

a Circle **one** word that is an erosion process.

b Underline **one** word that is a transportation process.

c Match **three** of the terms to the correct definition.

Coastal landscapes ⬭

longshore drift   solution   attrition   hydraulic power   slumping

Process	Description
	A sudden movement of soil and **unconsolidated** (loose) rocks down a cliff face.
	When pebbles bang into each other, making them smaller and more rounded.
	When waves forcefully throw water against a cliff face, causing some of the cliff to be worn away.

River landscapes ⬭

abrasion   hydraulic action   lateral erosion   vertical erosion   traction

Process	Description
	When pebbles are rolled along the bed of a stream by the force of the water flowing in the river.
	The sweeping motion of the water in a river, which widens each meander sideways.
	When stones are thrown against the banks of the river, causing some of the bank to be worn away.

Glacial landscapes ⬭

plucking   abrasion   rotational slip   freeze-thaw   bulldozing

Process	Description
	A weathering process. Rocks are broken up when water gets into cracks, freezes and pushes the rocks apart.
	The sandpaper effect of stones, carried in the ice, scraping against rocks, which deepens the U-shaped valley of a glacier.
	When rocks are pulled from the back wall of a **corrie** by the ice.

27 A great way to revise how landforms change over time is by drawing and labelling sketches.

a Sketch **one** of the following landforms in the box below:

waterfall   coastal arch   corrie

b Use at least **two** key terms to help explain how the landform changes over time. An example has been done for you.

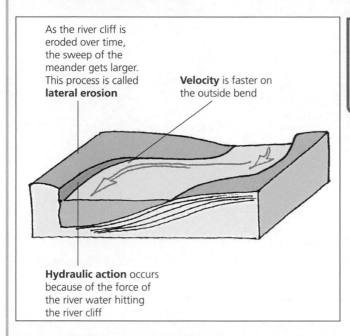

As the river cliff is eroded over time, the sweep of the meander gets larger. This process is called **lateral erosion**

Velocity is faster on the outside bend

Hydraulic action occurs because of the force of the river water hitting the river cliff

If the exam question is about a landform, you can include a sketch as part of your answer. The best sketches are carefully labelled to explain how geographical processes change the landform over time.

Figure 2.11 How a river meander changes over time

Examples of landforms

Use the table below to summarise key facts about landforms in each of the landscapes you have studied.

Landscape	Coast ⬭	River ⬭	Glacial ⬭	Coast ⬭	River ⬭	Glacial ⬭
Major landforms of erosion						
Major landforms of deposition						

Examples of management strategies

Use **two** tables to summarise key facts about management in the **two** landscapes you have studied.

Coastal landscapes – an example of a coastal management scheme

Location	
Why coastal management is needed	
How the management scheme works	
The effects of the management scheme (include conflicts)	

River landscapes – an example of a flood management scheme

Location	
Why flood management is needed	
How the management scheme works	
Social, economic and environmental issues	Social Economic Environmental

Glacial landscapes – an example of a glaciated upland area used for tourism

Location	
The attractions for tourists	1 2 3
Social, economic and environmental impacts of tourism	Social Economic Environmental
Strategies used to manage the impact of tourism	

Using OS maps

OS maps are often used in exam papers to assess your understanding of landscape. In order to answer questions, you need to be able to use:

- four-figure and six-figure grid references to locate places on the map
- the scale line and compass arrow to work out directions
- contour lines to understand the shape of the landscape.

OS maps are made at different scales. Two scales are often used in exam papers:

- 1:50,000 scale, where each 2 cm grid square represents 1 km² on the ground
- 1:25,000 scale, where each 4 cm grid square represents 1 km² on the ground.

> Always 'go along the hall before you go up the stairs' to get your grid reference numbers in the correct order.

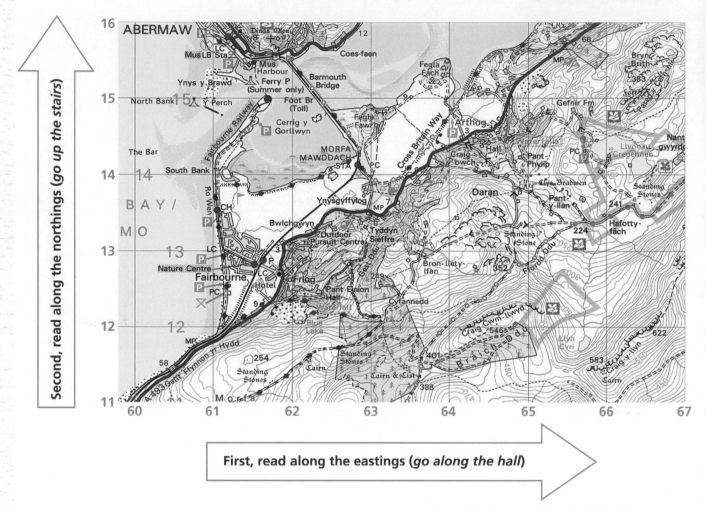

Figure 2.12 An OS extract at a scale of 1:50,000. © Crown copyright and database rights 2019 Hodder Education under licence to Ordnance Survey.

Four- and six-figure grid references

Four-figure grid references identify a complete grid square. Morfa Mawddach Station (shown with a red circle near the centre of the map) is in grid square 6214.

Six-figure grid references identify a specific location on the map. To turn a four-figure into a six-figure reference, you need to:

1 Find the third digit by estimating how many tenths of a kilometre the location is past the last easting. Morfa Mawddach Station is eight tenths past easting 62, so the first three digits are 628.

2 Find the sixth digit by estimating how many tenths of a kilometre the location is past the last northing. Morfa Mawddach Station is one tenth past northing 14, so the last three digits are 141.

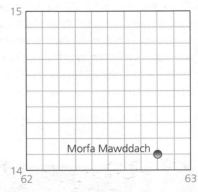

28 Match the following grid references to features on **Figure 2.12**.

6414   6511   6215   615128   658134

Grid reference	Feature
	Barmouth Bridge
	Lyn Cyri (a small lake)
	The village of Arthog
	A spot height of 224 metres
	Fairbourne Station

29 Which is the best description of grid square 6612?

a   An upland area with gentle slopes and woodland.

b   A steep-sided river valley. The river flows in a northerly direction.

c   An upland area with a steep slope which faces towards the northwest.

> Use the contour lines to imagine the shape of the land. The closer the contour lines, the steeper the gradient of the slope.

30 Match the four-figure grid references to the correct landform in the table below.

6012   6512   6511   6114   6611   6215

Grid reference	Landform
	Corrie lake (glacial)
	Arête (glacial)
	Coastal spit (coastal)
	Sandy beach (coastal)
	Estuary (river)
	V-shaped valley (river)

Using aerial photos with OS maps

Study **Figure 2.13**. It's an example of an aerial photo, taken by a drone. We have already seen that OS maps use contour lines to show the shape of the land. In your exam you could be shown an OS map and an aerial photo to test whether you can orientate the photo. This means finding landmarks on both the photo and the map and lining them up. The test is to see whether you can work out in which direction the camera was pointing when the photo was taken.

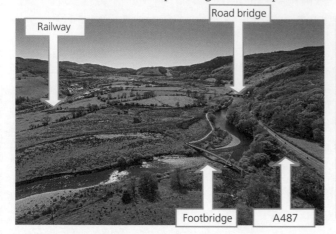

To work out the direction that the photo was taken in:

1 On the map, find at least two of the features that can be seen in **Figure 2.13**. You should be able to find the four features labelled.

2 Line up the features on the map as you can see them in the photo. Imagine you are a drone flying over **Figure 2.13**. You can see a road bridge behind the footbridge and the A487 road is to your right. To get these features lining up correctly, the drone must be in the top-right corner of the map over Glan Fechan farm in grid square 7502.

Figure 2.13 An aerial view of the River Dovey/Afon Dyfi

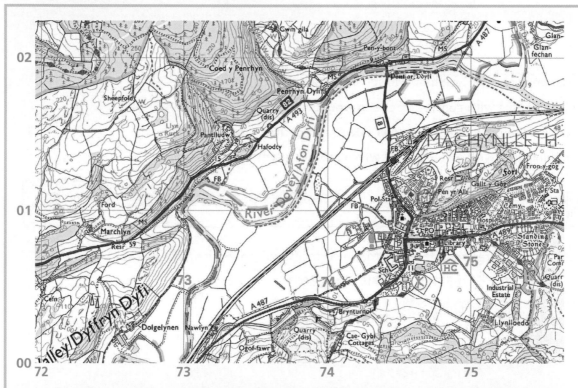

Figure 2.14 An OS extract at a scale of 1:25,000. © Crown copyright and database rights 2019 . Hodder Education under licence to Ordnance Survey.

31 Using **Figure 2.14**, give a six-figure grid reference for:

a the footbridge

b the road bridge

32 In which direction was the camera pointing?

Workbook answers can be found at **www.hoddereducation.co.uk/workbookanswers**

Use evidence in the photo

The trees on this bank are intercepting and storing water = **soft engineering**

The river is artificially straight so flood water moves quickly downstream. The concrete embankment increases the capacity of the channel = **hard engineering**

Figure 2.15 The River Ouse in York

33 Using Figure 2.15, <u>explain</u> the costs and benefits of strategies used to manage flood risk. **6 marks**

Use **evidence** in the photo. **3 marks for AO4 (skills)**

Show you **understand** how the strategies create advantages and disadvantages. **3 marks for AO2 (understanding)**

A good answer will be **balanced** between costs and benefits

The question uses a plural – so you need to explain more than one strategy

Use the **PEEL** technique for one hard engineering flood management strategy. Then use PEEL again for a second strategy. It would be good to use soft engineering for this second strategy to give your answer balance.

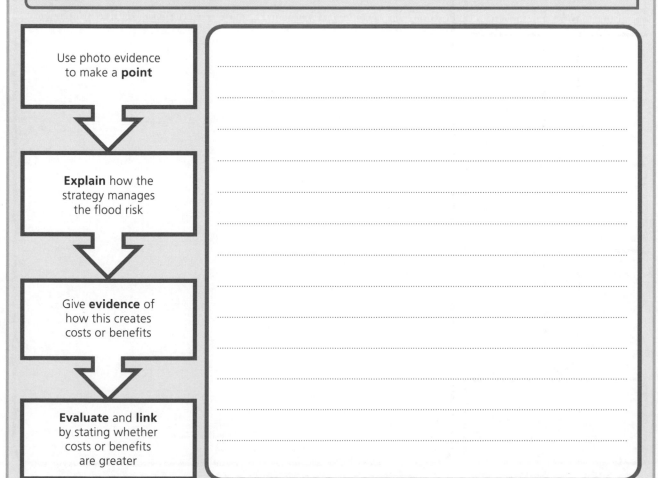

Use photo evidence to make a **point**

Explain how the strategy manages the flood risk

Give **evidence** of how this creates costs or benefits

Evaluate and **link** by stating whether costs or benefits are greater

34 'The benefits of coastal management outweigh the costs.' Do you agree? Using an example, explain your answer.

6 marks

Use evidence from an example of a coastal management scheme to back up your judgement. **3 marks for AO2 (understanding)**

Try to be balanced by writing as much about the advantages/benefits of coastal management as the disadvantages/costs

You must make a judgement. **3 marks for AO3 (justifying your decision)**

Use the PEEL technique to answer this question. Use evidence from an actual example of coastal management you have studied.

Make a **point** and …

Explain how the strategy works

Using an example, give **evidence** of how this creates costs or benefits

Evaluate and **link** by stating whether costs or benefits are greater

35 'The positive impacts of tourism in glacial landscapes outweigh the negative impacts.' Do you agree? Using an example, explain your answer. **6 marks**

You must make a judgement – this is **AO3**

Use evidence from an example of tourism in a glacial landscape you studied to back up your judgement

Try to be balanced by writing as much about the positive impacts of tourism (e.g. on the economy) as the negative impacts (e.g. on the environment)

Use the PEEL technique to answer this question. Use evidence from an actual example of tourism you have studied.

Make a **point** and …

Explain how tourism creates an impact on people, the economy or the environment

Using an example, give **evidence** of how this impact is positive or negative

Evaluate and **link** by stating whether positive or negative impacts are greater

Chapter 3: Preparing for Paper 2

Global urban patterns

Key abbreviations used in Paper 2 are **LIC**, **NEE** and **HIC**:

- **Low-income countries (LICs)** are very poor countries. Most LICs are in sub-Saharan Africa, such as Sierra Leone, Uganda and Tanzania. Towns and cities (**urban** places) are growing fastest in LICs.
- **Newly emerging economies (NEEs)** are developing countries which have a lot of industry. Many NEEs are in Asia (such as India and China) or South America (e.g. Brazil). The world's largest cities are found in NEEs – for example, Mumbai, Shanghai and Sao Paulo. Any city with over 10 million people is a **megacity**.
- **High-income countries (HICs)** are very wealthy countries such as the UK, Japan, Australia and the USA. There are some **megacities** in HICs, such as London, Tokyo and New York.

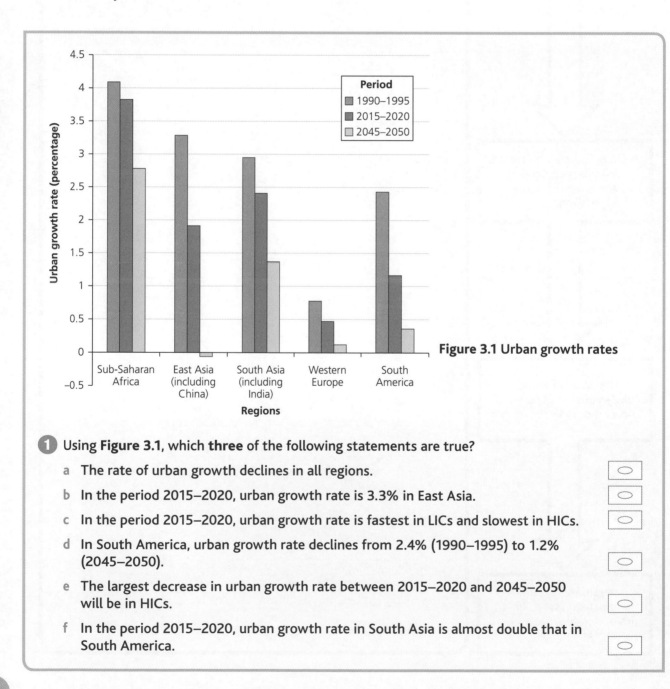

Figure 3.1 Urban growth rates

1 Using **Figure 3.1**, which **three** of the following statements are true?

a The rate of urban growth declines in all regions.

b In the period 2015–2020, urban growth rate is 3.3% in East Asia.

c In the period 2015–2020, urban growth rate is fastest in LICs and slowest in HICs.

d In South America, urban growth rate declines from 2.4% (1990–1995) to 1.2% (2045–2050).

e The largest decrease in urban growth rate between 2015–2020 and 2045–2050 will be in HICs.

f In the period 2015–2020, urban growth rate in South Asia is almost double that in South America.

Workbook answers can be found at **www.hoddereducation.co.uk/workbookanswers**

Push-pull theory

Urban areas in LICs and some NEEs continue to grow rapidly due to:

- **migration** – people moving there from the countryside (**rural** places)
- **natural increase** – more births than deaths.

Migration happens because of a mixture of:

- **push factors** – reasons you want to leave your home
- **pull factors** – reasons that attract you to move to somewhere else.

> Never give 'income' or 'jobs' as pull factors. People are pulled by the idea they might get a <u>higher</u> income or because there will be <u>more</u> jobs available.

2 Sort the following list into push factors and pull factors by placing each letter in the correct box.

A Shortage of food	**D** Higher incomes	**G** Flooding	**J** To join family
B Better pay	**E** Too far to nearest school	**H** To attend university	**K** Drought
C Lack of jobs	**F** Intolerance	**I** Conflict	**L** Desertification

Push factors

Pull factors

Urban opportunities in LICs and NEEs

Urban growth creates opportunities for people:

- **Social opportunities** – better access to schools, clean water and healthcare.
- **Economic opportunities** – large cities have better global links, for example international airports, so they have more industry.

3 a Explain how urban growth creates opportunities by linking the **points** on the left to the **explanations** on the right. One example has been done for you.

   b Use a highlighter to show how each explanation has been extended further.

Points	Explanations
Megacities have international airports and ports ...	... so there is a range of jobs in manufacturing with further opportunities for training and promotion for employees.
Universities are located in megacities ...	... so ordinary people have access to safe drinking water, which means there are fewer deaths at a young age.
Transnational companies build factories and offices in megacities ...	... so there are opportunities for firms that sell goods overseas to grow, which creates more jobs for local people.
Large cities in LICs and some NEEs continue to grow ...	... so there are opportunities for people to get degrees or vocational qualifications, which means they are more employable.
It is easier to provide water for an urban population than for people living in remote rural places ...	... so there is the opportunity to get healthcare quickly, which might prevent more serious health problems from developing.
You are never far away from a health clinic or hospital in a large city ...	... so there are always jobs available on construction sites requiring all sorts of skilled workers, from labourers to civil engineers.

> The 'So what?' technique has been used to explain each point (and highlighted in yellow). Notice how the sentence has been extended again to add further explanation (highlighted in green).

Urban challenges in LICs and NEEs

Urban growth creates social, economic and environmental challenges. For example, cities grow faster than housing can be built, resulting in the growth of **slums** and **squatter settlements**. Other challenges include reducing **unemployment**, dealing with **waste disposal** and managing **traffic congestion**.

---

4 Study the list of urban challenges below. Explain why **three** are a challenge. **Two** examples have been done for you.

A	Unemployment	**F**	Access to health services
B	Slum housing	**G**	Providing clean water
C	Traffic congestion	**H**	Sanitation
D	Air pollution	**I**	Reducing crime
E	Waste disposal	**J**	Water pollution

> Never write about just 'pollution'. It's too vague. Be precise by writing about 'air pollution' or 'water pollution'.

Traffic congestion is a challenge because it costs businesses a lot of money in wasted delivery time.

Water pollution is a challenge because a lot of plastic waste is dumped into urban rivers. This is washed into the ocean where it kills wildlife.

.... is a challenge because ....

....

....

....

.... is a challenge because ....

....

....

....

.... is a challenge because ....

....

....

....

Workbook answers can be found at **www.hoddereducation.co.uk/workbookanswers**

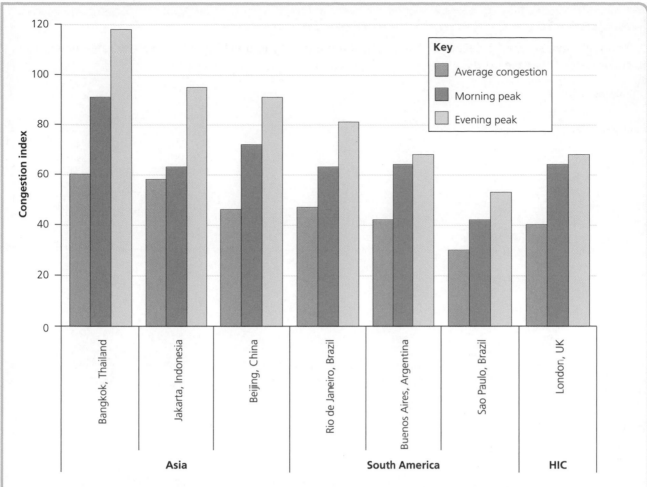

Figure 3.2 Congestion in selected megacities. The higher the index, the slower the traffic

5 Using **Figure 3.2**, what is the average congestion in London? ....

6 Using **Figure 3.2**, which **two** of the following statements are true?

a The most congested city in South America during the morning peak time is Rio de Janeiro. ⬭

b It is always more congested in the evening than in the morning. ⬭

c The average congestion index in Bangkok is 61 and the morning peak index is 95. ⬭

d Jakarta is busier than Beijing in the morning but Beijing is busier at night. ⬭

e Traffic congestion is a bigger problem in these Asian cities than in London. ⬭

A case study of a major city in an LIC or NEE

You will have studied a detailed case study of a major city in an LIC or NEE. Use the table below to summarise this case study.

Name of city		
Why the city has grown		Use bullet points to list key facts and figures. Be specific
Opportunities created by urban growth	Social	
	Economic	
Challenges created by urban growth	Social and economic	E.g. sanitation, health, slums, unemployment or crime
	Environmental	E.g. waste disposal, water pollution or traffic congestion

Workbook answers can be found at **www.hoddereducation.co.uk/workbookanswers**

Figure 3.3 Dharavi slum in Mumbai

Use evidence in the photo to make a point, then use 'So what?' technique to explain it

Make notes in the blank boxes above about how you can use the photo in **Figure 3.3**. Write your answer to this question in your notebook.

7 Using **Figure 3.3** and your own understanding, explain how urban growth creates social and environmental challenges.

6 marks

Use some evidence from the photo to support your answer. **3 marks for AO4 (skills)**

Show you understand, for example, by explaining why the growth of slums is a challenge – making links to crime, poor sanitation or water pollution. **3 marks for AO2 (understanding)**

An example of urban planning

The lives of the poorest people can be improved with better housing, **sanitation** or water supply or by providing health clinics. Some of these improvements are made by the people themselves – these are called **self-help projects**.

You will have studied an example of how urban planning can improve the lives of poor people. Use the table below to summarise your example.

Name of city	
Details of the urban planning scheme	E.g. improved sanitation or a self-help housing scheme
How this improved the lives of poor people	Use bullet points to list key facts and figures. Be specific

Urban change in the UK

This topic has a number of key geographical terms that could be used in exam questions. Learn them carefully.

| integrated transport | urban greening | deprivation | inequality | brownfield |
| greenfield | urban sprawl | commuter | migration | regeneration |

---

8 Match **six** key terms to the correct definition below. **Two** terms are not needed.

| brownfield | greenfield | regeneration | inequality |
| commuter | urban sprawl | urban greening | deprivation |

Term	Definition
	Someone who travels between their home and place of work.
	Planting trees or creating small parks within a town or city.
	Land that has never been built on before.
	Land that has been built on before and is now ready for redevelopment.
	The growth of a town or city outwards into the countryside.
	A lack of work, money, good health, a reasonable home or education.

Opportunities in UK cities

Change should be a good thing. Change in UK cities creates **opportunities** for people, businesses and the environment. For example, migration from other parts of the world gives UK cities a healthy **cultural mix**. As town centres are used less for shopping, they can be used for more **recreation** and **entertainment**. Planting trees and creating parks is a process called **urban greening**. Linking train stations to trams, bus routes and cycle paths is called **integrated transport**. For example, in **Figure 3.4**, this park links cycle paths to Cardiff Central train station. It is used by hundreds of commuters each day.

Figure 3.4 Cycle paths like this one allow commuters to travel safely through the city

9 Using Figure 3.4, **explain how** integrated transport creates benefits for people and the economy.

6 marks

You must use evidence from the photo. Think about how walking and cycling are good for your health. **3 marks for AO4 (skills)**

Show you understand, for example by explaining how better public transport means less traffic on the roads, which means less pollution in the air and less time stuck in traffic jams. **3 marks for AO2 (understanding)**

Don't forget this instruction. You should write one short paragraph about the benefits for people (e.g. health) and one short paragraph about the benefits for the economy (e.g. less money wasted because there is less congestion)

Write your answer to this question in your notebook.

Workbook answers can be found at **www.hoddereducation.co.uk/workbookanswers**

Challenges in UK cities

Urban change creates problems or **challenges**. Old **derelict** buildings can be expensive to demolish. The economic decline of some industries has led to social **deprivation** and **inequalities in housing**. Meanwhile, the growth at the edge of our cities – a process known as **urban sprawl** – means that **greenfield** sites are being built on. This has led to a rise in **commuting** and traffic **congestion**.

10 Study the list of urban challenges below. Explain why **three** are a challenge. **One** has been done for you.

A Social deprivation

B Inequalities in housing

C Dereliction

D Urban sprawl

E Growth of commuter settlements

F Building on brownfield sites

G Building on greenfield sites

H Waste disposal

The growth of commuter settlements is a challenge because *it increases demand for rail services used by commuters, which means rail companies need to spend a lot of money on extra services.*

.... is a challenge because ....

....

....

.... is a challenge because ....

....

....

.... is a challenge because ....

....

....

An example of urban regeneration

Urban regeneration projects aim to give new life to run-down urban areas. For example, port cities such as Liverpool and Cardiff have old dock areas that contained a lot of **dereliction**. By removing the dereliction and building a mixture of new housing, offices and recreation facilities, these cities have created better places for people to live and work.

You will have studied an example of urban regeneration in a UK city. Use the table below to summarise your example. Write your answer to this question in your notebook.

Place name	
Reasons why the area needed regeneration	1. 2.
The main features of the regeneration project	1. 2. 3.

List key facts and figures. Be specific

A case study of a major UK city

You will have studied a detailed case study of a major city in the UK. Use the table below to summarise this case study.

Name of city		
Location of the city		
Effects of migration		Use bullet points to list key facts and figures. Be specific
Opportunities created by urban change	Social and economic	E.g. mix of cultures, leisure, employment or integrated transport
	Environmental	E.g. urban greening
How urban change has created challenges	Social and economic	E.g. urban deprivation or inequalities in housing
	Environmental	E.g. waste disposal
	Impacts of urban sprawl	E.g. growth of commuting and building on greenfield sites

Sustainable urban living

The concept of **sustainability** means that planners try to make UK cities better places for people now and for future generations. Planners use various strategies to manage resources (such as **water** and **energy**) and improve **urban transport** to **reduce congestion**.

11 Match **five** strategies to the examples below. You may use some strategies more than once.

water conservation   energy conservation   waste recycling   urban greening   urban transport

Strategy	Example
	Create cycle lanes and bus lanes.
	Make householders sort their waste before it is collected.
	Make sure that all new houses have water meters.
	Add tramlines to existing city streets, e.g. in Nottingham.
	Create small **pocket parks** in inner-city areas.
	Introduce a **congestion charge** – a payment to drive in the city centre.
	Encourage householders to use a smart meter.

Figure 3.5 Urban greening in Cardiff. This park includes a cycle path that is used by hundreds of commuters each day

> You must use features shown in the photo to answer this kind of question. You could use the cycle path or the park in this example.

12 Using **Figure 3.5**, suggest how cities can be made more sustainable.

....

....

....

13 To what extent has an urban regeneration project overcome <u>economic and environmental</u> <u>challenges</u> in a UK city you have studied?

9 marks

To what extent is the command so … you must show you can evaluate/weigh up. **3 marks for AO3 (evaluation)**

You must use your knowledge of a case study. **3 marks for AO1 (facts)**

You must show you understand how regeneration improves the urban environment **and** the economy. **3 marks for AO2 (understanding)**

Answer this question in three short paragraphs. First, create an argument. Second, a counter-argument. Finally, use the 'washing line' technique.

On the one hand …

....

....

....

....

....

....

....

Use **PEEL** to make an argument that regeneration has been successful. Explain how it has overcome one economic and one environmental challenge

On the other hand …

....

....

....

....

....

....

....

Create a counter-argument. Use evidence to show why urban regeneration has not been successful

I fully/largely/partially agree/disagree …

....

....

....

....

Link back to the question. State whether urban regeneration has been successful or not. Use the 'washing line' technique

The changing economic world

There are a number of different ways that development can be measured. Some are measures of **economic development** while others measure social development (**quality of life**). You need to know:

- what is measured and what the measure tells us about the level of development
- the limitations of these measures.

A number of key geographical terms are important. Learn them carefully.

Gross National Income   **birth rate**   **death rate**   **infant mortality**
life expectancy   **people per doctor**   **literacy rate**   **access to safe water**
Human Development Index (HDI)

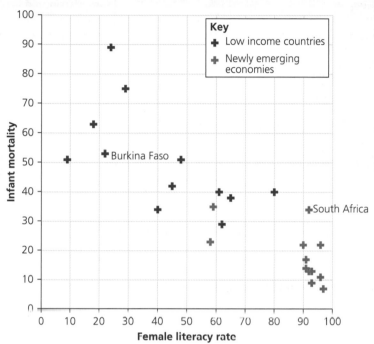

> To draw a line of best fit you should try to get an equal number of points on each side. It does not have to go through the origin (where the x-axis and the y-axis meet).

Figure 3.6 Scatter graph showing the correlation between female literacy rate and Infant mortality

14 Add a line of best fit to **Figure 3.6**.

15 Using **Figure 3.6**, which **three** of the following statements are true?

  a There is no correlation between the two sets of data.

  b The graph shows a negative correlation.

  c In South Africa, female literacy is 34 and infant mortality is 92.

  d In Burkina Faso, female literacy is 22 and infant mortality is 53.

  e NEEs tend to have lower female literacy and lower infant mortality than LICs.

  f NEEs tend to have higher female literacy and lower infant mortality than LICs.

16 Using **Figure 3.6**, explain why it is important for LICs to improve education for girls.

....

....

....

....

....

....

Figure 3.7 Map of Africa showing HDI

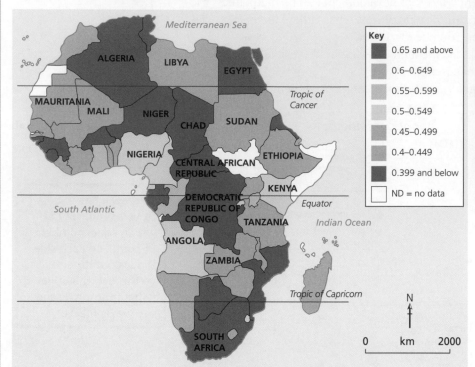

Figure 3.7 is a **choropleth** map. Countries have been put into categories depending on the value of their HDI. Use words like **clustered** or **scattered** and use the north arrow to describe how countries in the same category are distributed across the map.

17 Using **Figure 3.7**, name **three** countries that have an HDI of 0.399 and below.

....

18 Describe the distribution of countries with an HDI of 0.65 and above.

....

....

....

19 Of the 46 countries shown on **Figure 3.7**, what percentage has an HDI of 0.65 or above? Answer to 1 decimal place. Show your working.

The Demographic Transition Model

Birth rates and **death rates** gradually change as a country becomes more economically developed. As a result, the size and **structure** of the population (the number of older people compared with younger people) also change over time. These changes are predicted in the **Demographic Transition Model (DTM)**.

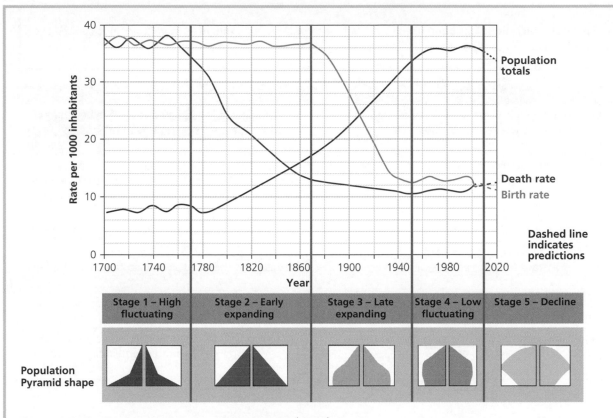

Figure 3.8 The Demographic Transition Model (DTM)

20 Using **Figure 3.8**, match the stages of the DTM to the correct descriptions.

Stage	Description
	Birth rate falls rapidly. The death rate is already quite low and continues to fall steadily.
	The population begins to fall as the death rate rises above the birth rate.
	Both the birth rate and the death rate are very high, so the population remains small.
	Both the birth rate and the death rate are low. The population is large but stops growing.
	The death rate falls rapidly but the birth rate remains high, so the population begins to grow.

21 Match **five** measures to the correct definition below. **Two** terms are not needed.

Gross National Income   birth rate   infant mortality   life expectancy
literacy rate   access to safe water   Human Development Index (HDI)

Term	Definition
	The average age someone can expect to live to.
	A measure of development that takes into account a country's level of education, its wealth and its average life expectancy.
	The number of children born in one year for every 1,000 people in a country's population.
	The number of children who die before the age of one for every 1,000 that are born.
	An economic measure that represents the average income in a country.

a Explain how the DTM is linked to the level of development by linking the **points** on the left to the **explanations** on the right. One example has been done for you.

b Use a highlighter to show how each explanation has been extended further.

> The 'So what?' technique has been used to explain each point (and highlighted in yellow). Notice how the sentence has been extended again to add further explanation (highlighted in green).

Points	Explanations
The status of women in most NEEs is improving, which means women have a more equal place in society ...	... so people live longer and have fewer children, which means that the population stops growing (stage 4).
Manufacturing and service industries have replaced agriculture as the main form of employment in NEEs ...	... so education for girls improves and better jobs become available, so women choose to have fewer children (stage 3).
In HICs, healthcare is good and many women choose to start their family later in life ...	... so deaths from cholera and typhoid are falling, so death rates are falling (stage 2).
In many LICs, pay is low and there are no social benefits or pensions for poor people ...	... so fewer people work on farms and there is much less need for children to work, so the birth rate falls (stage 3).
In most LICs, access to clean water is improving (although not everyone has access to it) ...	... so infant mortality is falling, which means that the death rate also falls (stage 3).
The use of mosquito nets in LICs is preventing the spread of malaria among vulnerable people ...	... so parents rely on their children to help the family earn a living, so birth rates remain high (stage 2).

Reducing the development gap

A number of **strategies** are used to create economic development in LICs and NEEs so that the **development gap** between rich and poor countries is reduced.

23 Match **six** strategies to the explanations below. One has been done for you.

Strategy	Point	Explanation
Intermediate technology	Some or all of the money owed by a country is cancelled or interest rates are lowered ...	... so that the country relies less on expensive imports of key goods.
Industrial development	Uses simple tools or machines ...	... so the government has more money available for development projects.
International aid	Investment is made in simple manufacturing such as steel ...	... so the poorest people can start their own small business.
Fairtrade	People who do not normally qualify for loans from a normal bank get a small loan ...	... so that they can be bought cheaply and maintained at low cost.
Debt relief	Provides money and expertise for a specific project, for example clean water ...	... so that money can be spent on community schemes like clinics or schools.
Microfinance	Buyers pay farmers a fair price for produce such as bananas and they also pay extra money ...	... so that local people have better health or education.

An example of tourism

Tourism is one strategy used by many LICs and NEEs to reduce poverty and create jobs. People are employed **directly** in new airports and hotels. Tourism also **indirectly** creates extra employment for people like farmers and fishermen who supply food to the hotels. However, many large hotels are owned by transnational companies (**TNCs**) so some profit is lost. Jobs created by tourism may be **seasonal** or badly paid.

You will have studied an example of how the growth of tourism in a LIC or NEE helps to reduce the development gap. Use the table below to summarise your example.

Place name	
What attracts tourists	
How tourism helps development	Use bullet points to list key facts and figures
Any limitations of this strategy	Reasons why tourism doesn't help development. Be specific

Transnational companies

Transnational companies (TNCs) are large businesses that have **branches** (factories, research facilities and offices) in more than one country. They have **advantages** for the **host** countries – for example, they create employment. There are also **disadvantages** – for example, at times of recession the TNC may close branches in host countries to protect jobs in the home country.

24 Add colours to the key below. Then use the key to colour the twelve statements.

Key		Advantages for the host		Disadvantages for the host
		Advantages for the TNC		Disadvantages for the TNC

1 Jobs are created at various levels of skill.	**5** Wages are often lower in the host country.	**9** Workers are trained and learn new industrial or business skills.
2 Local taxes may be lower than in the country where the TNC has its head office.	**6** Workers spend money in local shops.	**10** Environmental laws (e.g. about waste) may be less strong in the host country.
3 NEEs have large populations who are potential customers.	**7** The most highly paid jobs (e.g. research) stay in the country where the TNC has its head office.	**11** The TNC pays local taxes.
4 There may be no minimum wage.	**8** The local government is able to spend tax revenue.	**12** Other local firms get contracts, e.g. firms that supply component parts.

Figure 3.9 The location of Toyota's branches outside of Japan

25 How many vehicles are produced each year in Brazil? ....

26 Using **Figure 3.9**, outline **two** reasons why Toyota has branches in several NEEs. `2 marks`

....

....

> Some students write far too much for 2-mark questions – giving lots of reasons. Make sure you give **two** reasons. There are only 2 available marks so keep it simple and brief.

Workbook answers can be found at **www.hoddereducation.co.uk/workbookanswers**

27 **Assess the extent** to which **tourism** is able **to reduce the development gap.** `9 marks`

> You must show you can weigh up the evidence and reach a judgement. **3 marks for AO3 (evaluation)**

> Use your knowledge of tourism in an LIC or NEE. **3 marks for AO1 (facts)**

> You must show you understand how tourism can create jobs, reduce poverty and increase wealth. **3 marks for AO2 (understanding)**

Answer this question in three short paragraphs. First, create an argument. Second, a counter-argument. Finally, use the 'washing line' technique.

On the one hand ...

....

....

....

....

....

....

....

> Use **PEEL** to make an argument that tourism is able to create jobs – you could mention direct and indirect jobs

On the other hand ...

....

....

....

....

....

....

....

....

....

> Create a counter-argument. Use evidence to show why tourism can fail to reduce the development gap. You could mention the role of TNCs

I fully/largely/partially agree/disagree ...

....

....

....

> **Link** back to the question. State whether tourism is an effective strategy or not. Use the 'washing line' technique

A case study of one LIC or NEE

You will have studied a detailed case study of one LIC or NEE. Use the table below to summarise this case study. Use bullet points to list key facts and figures.

Name of LIC/NEE	
Why the country is important	
The changing economy	E.g. has there been a shift from farming to manufacturing?

Positive impacts of transnational companies	Negative impacts of transnational companies

Trading Partners	What is traded

Types of Aid	Impacts of aid
	E.g. water or health projects

Environmental impacts of development	E.g. pollution from industry or damage to ecosystems like rainforests
How development improves quality of life	E.g. improvements to education or getting rid of child labour

Economic futures in the UK

The UK economy is changing. Some **traditional industries**, such as coal mining and steel, have declined. Others, such as **service industries**, **finance** and **research**, have grown, leading to what some geographers call a **post-industrial economy**. These changes affect **employment patterns**, with some **regions** of the UK experiencing faster growth than other regions – leading to a **north–south divide**.

28 Match **five** key terms to the correct definition below. **Two** terms are not needed.

globalisation   de-industrialisation   post-industrial economy   science parks
infrastructure   airport/port capacity   north–south divide

Term	Definition
	The process that links the economies of different countries around the world.
	The amount of goods and people that can be moved through a port or airport.
	Locations where businesses set up their research operations.
	Structures such as roads, bridges and communication systems that are needed by the economy.
	The process that results in the decline of traditional manufacturing industry.

29 Study the list below. Use colours to complete the key for the causes and effects of de-industrialisation, then use the key to colour each statement.

Key	☐ Causes		☐ Effects

Decline of raw materials	Cheaper labour overseas	Need for re-training
Dereliction	Cheap imports	Regional inequalities
Unemployment	Lack of investment	

An example of modern industrial development

You will have studied an example of how modern industrial development can be more environmentally sustainable than previous types of industry. Use the table below to summarise your example. Use bullet points to list key facts and figures. Be specific.

Name of industry	
Where it is	
How the industry is more sustainable for the environment	1. 2.

The post-industrial economy

In the UK's **post-industrial economy**, businesses need good infrastructure, such as excellent road links and nearby airports. They also need a highly skilled workforce. The **south-east region** of the UK attracts a lot of **high-tech** companies. They benefit from being close to major **airports** (such as Heathrow), top-class **universities** (such as Reading or Cambridge) and **financial** advice in the City of London.

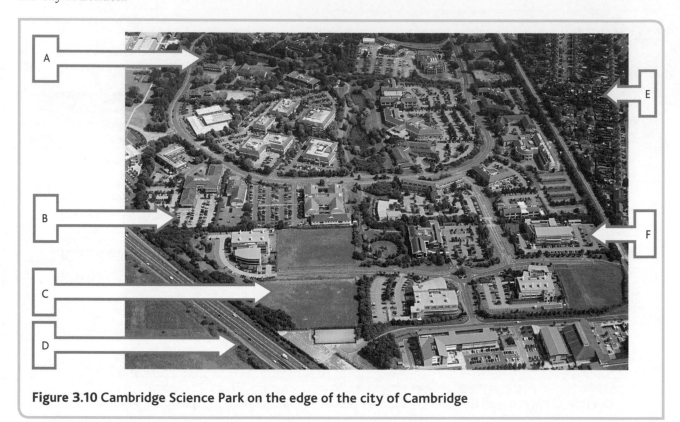

Figure 3.10 Cambridge Science Park on the edge of the city of Cambridge

30 Match each of the labels on **Figure 3.10** to the correct feature in the table below.

Label	Feature	Label	Feature
	Space for businesses to expand		Plenty of parking space
	Housing nearby for the workforce		Green, pleasant working environment
	Excellent road links to London		Other businesses nearby that can supply services

31 a   Explain how the features of the science park help businesses by linking the **points** on the left to the **explanations** on the right. One example has been done for you.

Points	Explanations
Other scientific and technology firms are located on the site ...	... so it's easy for staff of a high-tech TNC to visit branches abroad.
The Science Park is close to Cambridge University ...	... so businesses can share expertise and experiences.
Cambridge is only 40 minutes from Stansted Airport ...	... so businesses can recruit highly qualified graduate scientists and engineers.
The science park is next to a major road system ...	... so young graduate workers will be able to live close to their workplace.
Housing in Cambridge is cheaper than in London ...	... so components can be delivered quickly and efficiently.

Workbook answers can be found at **www.hoddereducation.co.uk/workbookanswers**

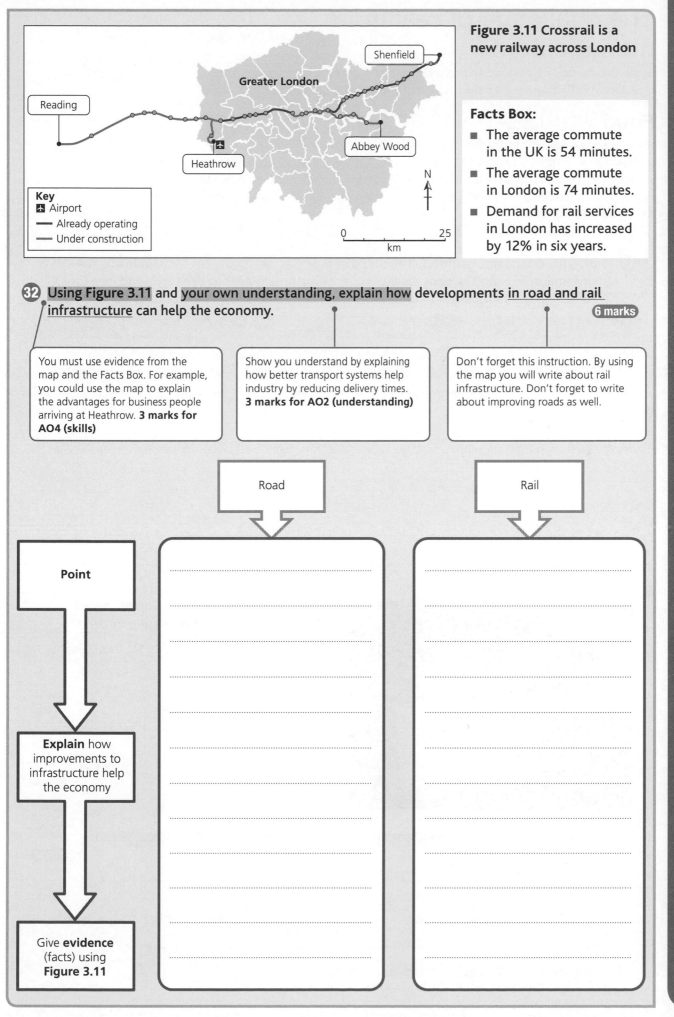

Figure 3.11 Crossrail is a new railway across London

Greater London

Reading

Shenfield

Heathrow

Abbey Wood

Key
✈ Airport
— Already operating
— Under construction

N

0   25
km

Facts Box:
- The average commute in the UK is 54 minutes.
- The average commute in London is 74 minutes.
- Demand for rail services in London has increased by 12% in six years.

32 **Using Figure 3.11** and your own understanding, explain how developments in road and rail infrastructure can help the economy. **(6 marks)**

You must use evidence from the map and the Facts Box. For example, you could use the map to explain the advantages for business people arriving at Heathrow. **3 marks for AO4 (skills)**

Show you understand by explaining how better transport systems help industry by reducing delivery times. **3 marks for AO2 (understanding)**

Don't forget this instruction. By using the map you will write about rail infrastructure. Don't forget to write about improving roads as well.

Road

Rail

Point

Explain how improvements to infrastructure help the economy

Give **evidence** (facts) using **Figure 3.11**

The challenge of resource management

Section C in Paper 2 assesses your understanding of resource management.
Question 3 is compulsory and you will revise that content on these two pages.

Food

Food is **seasonal**. For example, strawberries grow in the UK in June.
Consumers in the UK expect to be able to buy seasonal food at any time of year, so demand for food **imported from LICs** has grown. Importing food has a big impact on the environment because of **food miles** (the distance food travels) and its **carbon footprint**.

33 The demand for seasonal food at any time of year has consequences for people and the environment. Explain the consequences of growing and importing seasonal food into the UK from Kenya by adding explanations to the table below. **One** example has been done for you.

Points	Explanations	Further explanation
Crops such as beans and cut flowers are labour intensive ...	... so Kenyans are employed on farms	... which reduces poverty.
Farm land in Kenya that is used to grow beans could be used to grow crops to feed Kenyans ...	... so	
Beans and fresh flowers perish very quickly ...	... so	
Pesticides and insecticides are used to keep crops healthy ...	... so	

Figure 3.12 Fine beans in UK supermarkets are imported by air from Kenya during the UK's winter

34 Using **Figure 3.12**, discuss the issues arising from the demand for seasonal products at all times of the year.

6 marks

The command means you must consider the arguments that can be debated around a geographical issue. **3 marks for AO3 (analysis and evaluation)**

Make links that explain how our demand for seasonal products can create issues, for example food miles. **3 marks for AO2 (understanding)**

Write your answer to this question in your notebook.

Workbook answers can be found at **www.hoddereducation.co.uk/workbookanswers**

Water

Demand for water is increasing in the UK. Rainfall is highest in the west and north of the UK but demand for water is highest where the population is greater – in the Midlands and south-east. This means that some parts of the UK have a water **surplus** (more than enough) but other areas have a **deficit** (not enough). As a result, water has to be **transferred** from one region to another.

35 Match **five** key terms to the correct definition below. **Three** terms are not needed.

water surplus   water deficit   water transfer scheme   groundwater

abstraction   over-abstraction   aquifer   phosphates

Term	Definition
	Water stored in soil and in the rocks.
	Water trapped in porous rocks that can be used as a supply of water.
	Chemicals used in agriculture that can pollute water in the ground.
	The process of taking water from a river or from the ground.
	The use of rivers and pipes to move water from areas of water surplus to areas of water deficit.

Energy

The UK gets its energy from a mixture of **fossil fuels** (such as coal and gas), nuclear and **renewables**. This so-called **energy mix** is gradually changing because we now rely on coal much less than in the past.

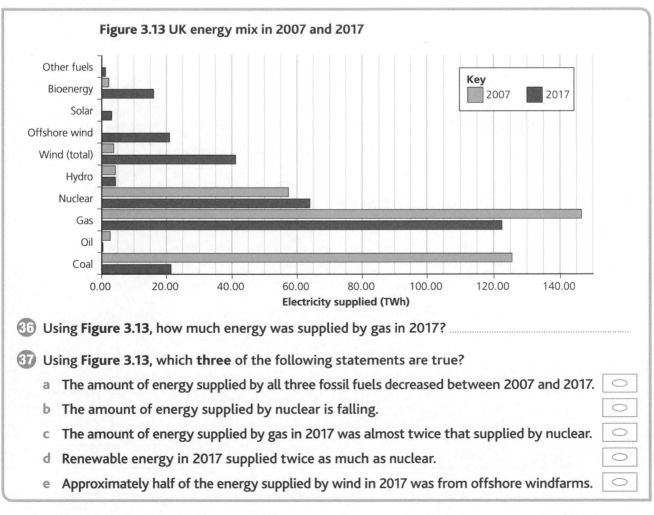

Figure 3.13 UK energy mix in 2007 and 2017

Key: 2007, 2017

Electricity supplied (TWh)

36 Using **Figure 3.13**, how much energy was supplied by gas in 2017? ....

37 Using **Figure 3.13**, which **three** of the following statements are true?

a The amount of energy supplied by all three fossil fuels decreased between 2007 and 2017.

b The amount of energy supplied by nuclear is falling.

c The amount of energy supplied by gas in 2017 was almost twice that supplied by nuclear.

d Renewable energy in 2017 supplied twice as much as nuclear.

e Approximately half of the energy supplied by wind in 2017 was from offshore windfarms.

Optional content on food/water/energy

As part of your study of resource management you will have concentrated on **one option** from food, water or energy. Tick the option you studied here:

Food ☐   Water ☐   Energy ☐

An example of a large-scale resource scheme

You will have studied a large-scale example of a food/water/energy scheme to show it can have both advantages and disadvantages.

Tick **one** example you studied:

A large-scale agricultural development ☐

A large-scale water transfer scheme ☐

The extraction of a fossil fuel ☐

Use the table below to summarise your example.

Location of the scheme	
How the scheme works	
Social or economic impacts	
Environmental impacts	

Impacts could be good or bad

An example of a local resource scheme

You will have studied an example of a small-scale food/water/energy scheme to show how it affects the local area.

Tick **one** example you studied:

A local scheme to increase food supplies ☐

A local scheme to increase water supplies ☐

A local renewable energy scheme ☐

Use the table below to summarise your example.

Location of the scheme	
How the scheme works	
Why the scheme is sustainable	

Explain how the scheme helps the environment and local incomes. Be specific

Tick **one** resource you have studied:

Food ⬭   Water ⬭   Energy ⬭

38 **Using an example** and your **own understanding**, explain how large-scale agricultural development can have <u>advantages and disadvantages</u>.   **6 marks**

You must use details from your example to support your answer. Be specific. **3 marks for AO1 (facts)**

Try to be balanced. Make sure you explain some of each

Make sure you make clear links between the large-scale scheme and its impacts. Use connectives like 'so' and 'which means that'. **3 marks for AO2 (understanding)**

39 **Using an example** and your **own understanding**, explain how a large-scale water transfer scheme can have <u>advantages and disadvantages</u>.   **6 marks**

40 **Using an example** and your **own understanding**, explain how a fossil fuel extraction scheme can have <u>advantages and disadvantages</u>.   **6 marks**

	Advantages	Disadvantages
Point		
Explain how large-scale scheme creates impacts		
Give **evidence** (facts) using your example		

Tick **one** resource you have studied:

Food ⬭   Water ⬭   Energy ⬭

41 'Local schemes are an <u>effective</u> way to increase sustainable supplies of food.' Do you agree?
Using an example, explain your answer.   6 marks

> Make sure you make clear links between a local scheme and how it increases supplies. Use connectives like 'so' and 'which means that'. **3 marks for AO2 (understanding)**

> Is the scheme effective or not? Think about the effects of the scheme on the people and the environment. Perhaps it is only *partially* successful.

> This command means you must weigh up the evidence and come to a decision. **3 marks for AO3 (evaluation)**

42 'Local schemes are an <u>effective</u> way to increase sustainable supplies of water.' Do you agree?
Using an example, explain your answer.   6 marks

43 'Local schemes are an <u>effective</u> way to increase sustainable supplies of energy.' Do you agree?
Using an example, explain your answer.   6 marks

Point

⬇

Explain how local scheme creates impacts

⬇

Give **evidence** (facts) using your example

Chapter 4: Preparing for Paper 3

Section A: The Issue Evaluation

Paper 3 has **two** Sections. Section A is the **Issue Evaluation**. Section A has a total of 34 marks and you must answer **all** of the questions. Section A finishes with a 9-mark question (with 3 extra marks for SPaG). This question tests how well you can **make a decision**.

> You should spend about 35 minutes in total on Section A.

Using the Pre-release resources booklet

The Pre-release resources booklet is sent to your school twelve weeks before the exam. This booklet is about eight pages long. It is full of information about a geographical issue. The information could take the form of short news articles, diagrams, maps, graphs, photos or people's points of view (speech bubbles). The geographical issue could focus on anything in the compulsory parts of Papers 1 and 2.

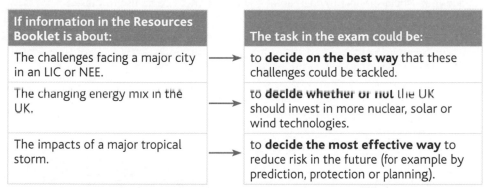

If information in the Resources Booklet is about:	The task in the exam could be:
The challenges facing a major city in an LIC or NEE.	to **decide on the best way** that these challenges could be tackled.
The changing energy mix in the UK.	to **decide whether or not** the UK should invest in more nuclear, solar or wind technologies.
The impacts of a major tropical storm.	to **decide the most effective way** to reduce risk in the future (for example by prediction, protection or planning).

In the Issue Evaluation the examiners want to see whether you can make sense of information that is presented to you in the Pre-release resources booklet. You should study the resources carefully in the weeks before the exam. To make your study of the resource booklet effective you need to find ways to be engaged. For example, you can:

- make notes all over your copy of the booklet. You will be given a new, clean copy of the resource booklet at the start of the exam
- highlight words or phrases in the booklet. For example, if the booklet contains a passage that describes a strategy to manage the issue:
  - use one colour pen to highlight any benefits/advantages of the strategy
  - use a *different* colour to highlight any negative impacts/disadvantages of the strategy.

> Make lots of notes on your copy of the Pre-release resources booklet. Then use your notes to complete **Figure 4.1**. You need to **revise** this because you can't take your own copy of the booklet into the exam with you.

Figure 4.1 My notes about the Pre-release resources booklet

What the information is about		
What the decision could be about		
Points of view in the booklet	For the issue	Against the issue
What would I decide?		
What evidence (from the booklet) could I use to support my decision?	Evidence that would support my decision	Evidence that I could use to make a counter-argument

Types of question in the Issue Evaluation

It's important to realise that none of the 34 marks in Section A is for remembering facts (AO1) from your examples or case studies. The examiners will use a variety of low-tariff questions (1, 2 or 3 marks) as well as **at least one** 6-mark question and **one** 9-mark question in the Issue Evaluation. The low-tariff questions will be simple:

- Some will assess whether you can use the maps or graphs in the resources booklet. These are questions that assess skills (AO4). See pages 6–10 for more advice. Questions could ask you to:
  - give a value from a graph
  - describe the relief shown on an OS map
  - describe a location or a distribution pattern.
- Other questions will assess your understanding (AO2) by using commands like:
  - Give **one** reason for …
  - Outline **two** effects of …

The main focus of the 6-mark and 9-mark questions is to assess your ability to **analyse** and **evaluate the information** in the resources booklet and/or your ability to **make a decision** (AO3).

> The focus on AO3 in Part A is important. It means you need to revise for this paper in a different way to Papers 1 and 2. There's no point trying to learn a lot of facts about the examples and case studies. Instead, you should spend time studying the resources in the booklet very carefully because you will be asked to **evaluate** them.

Workbook answers can be found at **www.hoddereducation.co.uk/workbookanswers**

Answering the 9-mark question

The 9-mark question in Paper 3 Section A is rather different to 9-mark questions you have come across before. In Papers 1 and 2 these questions assess a mixture of AOs (see pages 18–22). However, in Paper 3 the 9-mark questions (**one** in Section A and **one** in Section B) assess only your ability to **analyse** and **evaluate evidence** or **make a decision** (AO3).

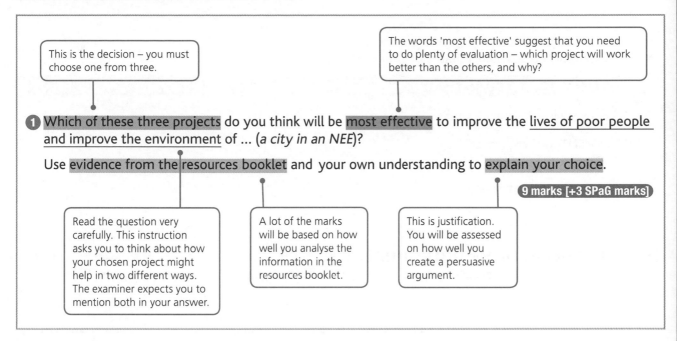

This is the decision – you must choose one from three.

The words 'most effective' suggest that you need to do plenty of evaluation – which project will work better than the others, and why?

1 Which of these three projects do you think will be most effective to improve the lives of poor people and improve the environment of … (*a city in an NEE*)?

Use evidence from the resources booklet and your own understanding to explain your choice.

9 marks [+3 SPaG marks]

Read the question very carefully. This instruction asks you to think about how your chosen project might help in two different ways. The examiner expects you to mention both in your answer.

A lot of the marks will be based on how well you analyse the information in the resources booklet.

This is justification. You will be assessed on how well you create a persuasive argument.

Figure 4.2 Breaking down the 9-mark questions in Paper 3, Section A

1 Use a highlighter pen or underlining to break down the following question. Write some labels to remind you what each part of the question means. Use **Figure 4.2** to help you.

Do you think the proposed development … (of a wind farm) … should go ahead?

Use evidence from the resources booklet and your own understanding to explain your decision.

9 marks [+ 3 SPaG marks]

Planning your answer to a 9-mark question

There is a total of 12 marks (9 plus 3 for SPaG) available, and you have about 13 minutes to complete your answer. You might be tempted to start writing straightaway, but you ought to plan your answer first. You could use a mind-map or a simple table like the one shown in **Figure 4.3**.

> Two minutes writing a plan is time well spent. Keep glancing back at your plan – and the question – as you write your full answer.

Figure 4.3 When you are planning your answer, use a table to evaluate the possible impacts of your decision

	Advantages of my decision	Disadvantages of my decision
SOCIAL (for people and communities)		
ENVIRONMENTAL (think about ecosystems, air pollution or water use)		
ECONOMIC (jobs and businesses)		

Structuring your 9-mark answer

Figure 4.4 shows how you might structure your answer in three simple paragraphs.

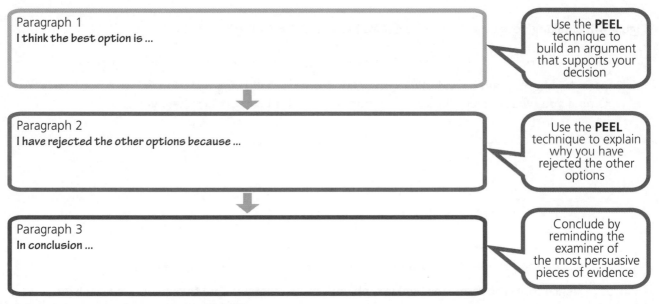

Paragraph 1
I think the best option is ...
→ Use the **PEEL** technique to build an argument that supports your decision

Paragraph 2
I have rejected the other options because ...
→ Use the **PEEL** technique to explain why you have rejected the other options

Paragraph 3
In conclusion ...
→ Conclude by reminding the examiner of the most persuasive pieces of evidence

Figure 4.4 How to structure your decision

It's very important that you explain why your decision is best. Do this in the first paragraph. You might realise that your decision could have some negative implications. That's okay, but it's important to explain why the advantages of your plan outweigh any disadvantages.

In the second paragraph, explain why the other options are not as good as the one you have chosen. It's a good idea to give a balanced account, so you should write about the possible advantages of these options before explaining why, on the whole, you have rejected them.

Finish your answer with a short conclusion. **Figure 4.5** gives you some advice about this final paragraph.

Figure 4.5 Dos and don'ts of writing a conclusion

Do	Don't
Do remind the examiner that you have looked at both sides of the argument.	Don't sit on the fence! If you have been asked to make a judgement, you should make it.
Do repeat what you think is the most persuasive or significant piece of evidence that supports your decision.	Don't state that it is difficult to make a decision if the earlier part of your essay is packed full of strong evidence.
Do use key words or phrases from the question in your conclusion.	Don't forget to glance back at the question before you start the conclusion. If you have wandered off task then now is your chance to save your essay and answer the question!

> If you use the word 'sustainable' make sure you explain why your option is sustainable.

Signposting

Examiners like the 6-mark and 9-mark answers to be structured, so a good answer will be organised into paragraphs. You can also use signposting to help structure your answer. Signposting is a technique that tells the reader what is coming next – like a signpost on the roadside tells you where you are going. **Figure 4.6** has a few useful signposts you can use when writing your longer answers.

Figure 4.6 Signposts for writing longer answers

Adding points	Writing in lists	Adding emphasis	Linking cause and effect
Moreover	Next	Especially	Because
Furthermore	Then	Particularly	Consequently
In addition	Lastly	Chiefly	Despite this
What is more	Firstly	Mainly	Therefore
	Secondly	Mostly	
	Finally		

Aiming high!

Your decision will have impacts and your answer must show that you understand this. Evaluating the possible social, economic and environmental impacts (page 84) is good. However, if you are aiming for a really high mark then you should go beyond this and evaluate one or more of the following:

- Possible consequences of your decision on **different** groups of people.
- Likely **spatial** impacts of your decision – local, regional or national.
- How your plan could have **short-term** and **long-term impacts**.

To help with this, think about the consequences of the damage to the railway track in **Figure 4.7**. A good candidate might use this resource to suggest that, until the line is repaired, commuters will have to use a bus service for part of their journey and this would be a significant inconvenience. They might also suggest that the consequences were local (repairs to the sea wall to make local properties safe) and national (Network Rail had to spend millions to repair the track – money that could not be spent elsewhere).

> When suggesting possible impacts you should avoid extreme statements. For example, using **Figure 4.7**, it would be extreme to state that commuters wouldn't be able to get to work and that they would lose their jobs!

Figure 4.7 Storm damage to the railway line at Dawlish in Devon

2 Study **Figure 4.7**. This railway line joins Cornwall to London. It was damaged in February 2014. It took two months and £35 million to make the repairs. Outline the possible impacts on each of the following:

a Three different groups of people

b The local area (Dawlish) and the region of Cornwall

> Think about impacts on local residents and businesses in Cornwall

c One short-term (during February–March 2014) impact and one longer-term (after April 2014) impact

> Think about how long it might take business to recover lost trade

Workbook answers can be found at **www.hoddereducation.co.uk/workbookanswers**

Section B: Fieldwork

Section B of Paper 3 assesses the six stages of **Fieldwork** shown in **Figure 4.8**. You must answer **all** of the questions in Section B. Section B has a total of 36 marks. None of these is for AO1 (remembering facts) or AO2 (understanding). 12 marks are for **geographical skills** (AO4) and a whopping **24 marks** are for **analysis**, **evaluation** and **decision making** (AO3).

> You should spend about 40 minutes in total on Section B.

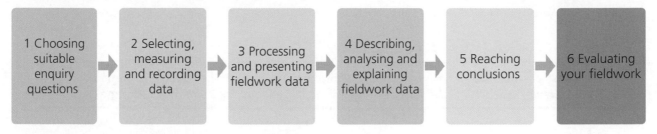

Figure 4.8 The six stages of the fieldwork enquiry

Questions about unfamiliar fieldwork

The first few questions in Section B are about **unfamiliar** fieldwork. This means that the questions give you some information such as a photograph, table of data, graph or map about fieldwork which has been done by someone else. Most of these questions are low tariff (1, 2, 3 or 4 marks). They may assess AO3 or AO4. Some important key words are used in fieldwork questions. Make sure you know what they mean:

accuracy   **reliability**   **suitable or appropriate**

Figure 4.9 A busy shopping street in central Birmingham

> Question 3 assesses AO3 – can you analyse the evidence in the photo? The street is busy and people are moving in different directions. The important word in the question is *accurate* – which means the results being as close to the real value as possible.

3 Using **Figure 4.9**, give **one** reason why it would be difficult to collect accurate data about the number of pedestrians using this street.　**2 marks**

....

....

....

....

4 A group of students counted the number of pedestrians at eight locations across the city centre. Explain what the students would need to do to make the results reliable. **2 marks**

....

....

....

....

> Question 4 also assesses AO3 – can you evaluate what the students did? The important word in the question is *reliable*. Reliability is about data being collected in exactly the same way each time.

Figure 4.10 A bipolar survey

Positive	+3	+2	+1	0	−1	−2	−3	Negative
Pleasant environment for pedestrians								Unpleasant environment for pedestrians

The students used a bipolar survey (shown in **Figure 4.10**) to score the quality of the environment at each of the eight locations. The students then chose different techniques to present the bipolar data. These are shown in **Figures 4.11** and **4.12**.

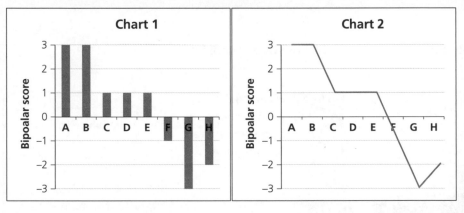

Figure 4.11 Bipolar score charts

5 Which is the most appropriate method in **Figure 4.11** for presenting the data shown? Give a reason for your choice. **2 marks**

....

....

....

....

> This question assesses geographical skills (AO4). The key word here is *appropriate*. Bar charts are appropriate for presenting **discrete data** (things that are counted or put into categories). Line graphs are appropriate for presenting **continuous data** (things that can be measured like height, time or velocity).

Workbook answers can be found at **www.hoddereducation.co.uk/workbookanswers**

One student chose to present the bipolar data as a map. It is shown in **Figure 4.12**.

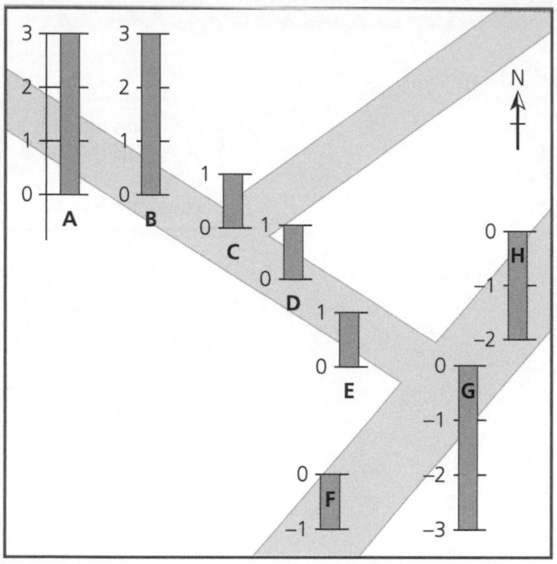

Figure 4.12 Bipolar data as a map

6 Describe the pattern of bipolar scores shown in **Figure 4.12**. `2 marks`

....

....

....

> This question assesses geographical skills (AO4). Use the north arrow. Never write about the 'top of the map' or the 'right of the map'

7 Suggest one way **Figure 4.12** could be improved. `1 mark`

....

....

....

> In this kind of question, always check that the map has a north arrow, key and scale line.

Questions about your own fieldwork

The remaining questions in Section B assess **your own fieldwork**. These questions mainly assess your ability to **evaluate your fieldwork** (AO3). They will include a 6-mark and a 9-mark question (with 3 extra marks for SPaG). This question tests how well you can make a **decision**.

> When revising you must **think critically** about your fieldwork and **evaluate** what you did because lots of questions assess AO3.

You will have carried out two fieldwork enquiries, one which focused on **physical** geography (such as fieldwork on a river or on a beach), the other which focused on **human** geography (such as urban geography). You need to remember what you did in your fieldwork – you should use **Figure 4.13** to make some revision notes. With so many marks available for evaluation (AO3) it is essential that you also consider:

- **why** you chose to do each stage of the fieldwork in that way
- the **strengths and weaknesses** of each stage of your fieldwork.

Figure 4.13 The six strands (or stages) of the fieldwork enquiry

Strands	Physical enquiry	Human enquiry
1 Choosing suitable enquiry questions		What was the aim of your fieldwork? Was it sensible and achievable?
2 Selecting, measuring and recording data		What sampling strategy did you use? Can you justify its choice? Should you have collected any other data?
3 Processing and presenting fieldwork data		What maps and graphs did you use? What were their strengths and limitations?
4 Describing, analysing and explaining fieldwork data		Did the data allow you to identify trends and patterns? If not, why not?
5 Reaching conclusions		Was the data reliable enough so that you could reach firm conclusions?
6 Evaluating your fieldwork		What might have happened if you had collected your data in a slightly different way or at a different time of day?

8 Complete **Figure 4.13** with key details about each of your geography fieldwork enquiries. Use bullet points.

Workbook answers can be found at **www.hoddereducation.co.uk/workbookanswers**

Command words

Command words in this section of the exam paper are used to make you justify or evaluate – although, strangely, the word evaluate is not used! Instead you will get questions that ask about strengths, limitations or why your fieldwork was effective, helpful or useful. Here are some examples which have been broken down.

1 **Justify** the choice of location for your human geography enquiry.   `4 marks`

> 'Justify' means that you must give reasons why you took this decision. In this case, you need to explain the advantages of using this location. You could consider access to the site or its safety.

2 **Explain** why one primary data collection technique used in your human geography enquiry was **effective**.   `3 marks`

> The command is 'explain', but you have to join it up with 'effective' to make sense of the question. The question is really asking you to evaluate data collection – and focus on the positive aspects.

> Follow the instruction carefully! Only do one. Make sure it was primary data (data collected by you) rather than secondary data. The examiner could set a question about secondary data too, of course!

3 **Explain** how the data presentation techniques used in your physical geography enquiry **helped you to analyse** the data.   `6 marks`

> The command is 'explain', but you have to join it up with 'helped you to analyse' to make sense of the question. The question is really asking you to evaluate your data presentation techniques – the maps, graphs or statistical tests you used.

> This question is worth 6 marks. You should be balanced – explaining why some techniques were helpful and why some weren't very helpful.

4 For **one** of your fieldwork enquiries, **to what extent** were your conclusions accurate and reliable?   `9 marks [+3 SPaG]`

> If you are asked to write about one fieldwork enquiry, don't waste time and effort writing about both. You won't get any extra marks!

> The command is 'to what extent', so you are being asked to make a judgement. Use the 'washing line' technique. Support your answer with evidence of how your data was and was not accurate and reliable.

> A really good answer will evaluate the whole fieldwork process. Did the choice of aims and methods of data collection provide you with enough data that was **accurate** and **reliable**? These are key terms – make sure you know what they mean.

Evaluation, evaluation, evaluation

We have seen that Section B has 36 marks and 24 of these are for AO3. You must be able to write a good evaluation of your fieldwork if you want high marks. This isn't easy. Students tend to fall into one of two traps when they are trying to evaluate their fieldwork:

Pitfall one: Some students write about what went well or what went badly, but this isn't necessarily good evaluation. For example, 'I didn't collect much data because it rained' isn't a good evaluation – it means you should have worn a better coat!

Pitfall two: A lot of students make rather vague statements that imply evaluation. For example, 'It was useful to have data from several different sites around the town'. The word 'useful' implies that this was a strength. It would be better to be much more direct. For example, 'It was a **major strength** of my sampling to collect data from several different sites **because** this allowed me to draw a conclusion about the spatial pattern in the data.' This response is much better because 'major' is a qualifying word – it tells the examiner *how* useful the data was. The student then goes on to explain *why* this was useful.

Use words from the lists in **Figure 4.14** to make sure you are actually signposting your evaluation in a really obvious way. Then the examiner will spot it and, hopefully, give you credit for it.

Figure 4.14 Words to use when evaluating your fieldwork

Words that describe a positive aspect of your fieldwork	Words that describe a negative aspect of your fieldwork	Qualifying words (adjectives)
Strength	Limitation	Significant
Advantage	Disadvantage	Substantial
Benefit	Weakness	Serious
Opportunity	Obstacle	Major
Merit	Challenge	Minor
Success	Failure	Partial

> Use these words to indicate relative importance, for example, 'minor limitation' or 'major strength'

Using your own experience

Lots of students attempt an evaluation, but it is very limited because it is so general – it could be about any piece of fieldwork. For example, 'we didn't really have a big enough sample'. Adding specific details about **your own experience** is important when you are evaluating your fieldwork. **Figures 4.15** and **4.16** give you some ideas about how to do this.

Figure 4.15 How to improve your evaluation by making it specific

> The evaluation is qualified

What you did	Limited evaluation	Specific evaluation
I used an OS map to help choose locations to sample data in the town.	OS maps are sometimes out of date.	A significant weakness of my sampling was that the Willowbridge housing estate was so new it wasn't on the OS map.
I used satellite images to choose places along the river where we could collect data.	Satellite images are useful because they are up to date.	A major strength of this was that I could see from the satellite image and overlay map where there was safe public access to the River Clun.

> The weakness is explained using specific detail

Figure 4.16 The difference between limited evaluations and specific evaluations

Tick which evaluation is better!

Data collection	Evaluation	
I measured the length of pebbles along a transect (a line) going up the beach profile.	The tide was coming in so I had to rush.	☐
	I didn't check the tide timetable and the tide was coming in. This was a major weakness because I couldn't take pebble sizes at the bottom of the beach profile so my transect is incomplete.	☐
I intended to use systematic sampling to ask a questionnaire of every tenth person.	Some people didn't want to answer our questions so we just asked people who looked friendly.	☐
	Some people were too busy to stop and talk so we had to ditch our systematic method and just ask anyone who would stop. This is a problem because our results were probably not representative.	☐
I measured wind speeds in the sand dunes every 20 minutes.	This was too infrequent because the wind was very gusty on the day we visited because a front had passed over.	☐
	However, the wind speed varied a lot in between the readings.	☐
I intended to do an Environmental Quality Index (EQI) every 100 m along a straight line from the suburbs to the city centre.	We couldn't keep to a straight line because there was a main road in the way. This was a minor limitation because it meant our sampling wasn't perfectly systematic.	☐
	It was difficult to keep to a straight line and do the readings at exactly 100 m.	☐
I worked as part of a large group collecting data on the amount of traffic all over the town.	One group of students collected their results 10 minutes later than everyone else so we cannot be sure that the cars they counted hadn't already been counted elsewhere.	☐
	We had problems getting everyone to collect the results at exactly the same time.	☐

9 a Read each of the evaluations in **Figure 4.16**. For each example of data collection, tick the evaluation that you think is better.

b Use a pen to circle where the evaluation is qualified.

c Use a highlighter pen to highlight part of the answer that provides **specific details** of the student's own experience.

10 To what extent were the primary data collection techniques used in either of your fieldwork enquiries effective?

6 marks

Be direct by using words like strength or limitation. Qualify it

Support your evaluation with some specific details

Evaluate a technique that worked well

Evaluate a technique that didn't work so well

Workbook answers can be found at **www.hoddereducation.co.uk/workbookanswers**

Notes

1 **This workbook will help you** prepare for your AQA GCSE (9–1) Geography exams.

2 **Build your skills** and prepare for every question in the exams using:
- clear explanations of what each question requires
- short answer activities that build up to exam-style questions
- spaces for you to write or plan your answers.

3 **Answering the questions** will help you build your skills and meet the assessment objectives (AOs):

AO1: remembering geographical facts

AO2: understanding geographical concepts and processes

AO3: evaluating evidence or using evidence to make a decision

AO4: using skills to investigate maps and graphs or making calculations.

4 **You still need to** read your textbook and refer to your revision guide and lesson notes.

5 **Answers** to every question in the book are available at: **www.hoddereducation.co.uk/workbookanswers**

The Publishers would like to thank the following for permission to reproduce copyright material.

Photo credits

p.11 © Andy Owen; **p.13** © RZAF_Images/Alamy Stock Photo; **p.15** © Andy Owen; **p.31** © Rainier Martin Ampongan/Alamy Stock Photo; **p.39** © Andy Owen; **p.41** © Hemis/Alamy Stock Photo; **p.50** © Keith Morris Aerial Imagery/Alamy Stock Photo; **p.51** © Andy Owen; **p.59** © RichSTOCK/Alamy Stock Photo; **p.60** © Andy Owen; **p.63** © Andy Owen; **p.74** © Cambridge Aerial Photography/Alamy Stock Photo; **p.76** © Andy Owen; **p.86** © Lightworks Media/Alamy Stock Photo; **p.87** © Andy Owen.

Acknowledgements

Map and data product images by kind permission of Ordnance Survey OS Explorer(s).

Every effort has been made to trace all copyright holders, but if any have been inadvertently overlooked, the Publishers will be pleased to make the necessary arrangements at the first opportunity.

Although every effort has been made to ensure that website addresses are correct at time of going to press, Hodder Education cannot be held responsible for the content of any website mentioned in this book. It is sometimes possible to find a relocated web page by typing in the address of the home page for a website in the URL window of your browser.

Hachette UK's policy is to use papers that are natural, renewable and recyclable products and made from wood grown in well-managed forests and other controlled sources. The logging and manufacturing processes are expected to conform to the environmental regulations of the country of origin.

Orders: please contact Hachette UK Distribution, Hely Hutchinson Centre, Milton Road, Didcot, Oxfordshire, OX11 7HH.
Telephone: +44 (0)1235 827827. Email education@hachette.co.uk
Lines are open from 9 a.m. to 5 p.m., Monday to Friday. You can also order through our website: www.hoddereducation.co.uk

ISBN: 9781510453364

© Andy Owen 2019

First published in 2019 by
Hodder Education,
An Hachette UK Company
Carmelite House
50 Victoria Embankment
London EC4Y 0DZ

www.hoddereducation.co.uk

Impression number 10 9 8

Year   2024

Cover photo © Frederic Bon – Fotolia.com

Illustrations by Aptara.

Typeset in India by Aptara.

Printed in Dubai.

A catalogue record for this title is available from the British Library.

With special thanks to:

Ordnance Survey

The world's trusted geospatial partner

HODDER EDUCATION
t: 01235 827827
e: education@hachette.co.uk
w: hoddereducation.co.uk

ISBN 978-1-5104-5336-4

9 781510 453364

MIX
Paper | Supporting responsible forestry
FSC™ C104740